Algorithms and Subjectivity

In this thought-provoking volume, Eran Fisher interrogates the relationship between algorithms as epistemic devices and modern notions of subjectivity.

Over the past few decades, as the instrumentalization of algorithms has created knowledge that informs our decisions, preferences, tastes, and actions, and the very sense of who we are, they have also undercut, and arguably undermined, the Enlightenment-era ideal of the subject. Fisher finds that as algorithms enable a reality in which knowledge is created by circumventing the participation of the self, they also challenge contemporary notions of subjectivity.

Through four case-studies, this book provides an empirical and theoretical investigation of this transformation, analyzing how algorithmic knowledge differs from the ideas of critical knowledge which emerged during modernity – Fisher argues that algorithms create a new type of knowledge, which in turn changes our fundamental sense of self and our concept of subjectivity.

This book will make a timely contribution to the social study of algorithms and will prove especially valuable for scholars working at the intersections of media and communication studies, internet studies, information studies, the sociology of technology, the philosophy of technology, and science and technology studies.

Eran Fisher is Associate Professor at the Department of Sociology, Political Science, and Communication at the Open University of Israel. He studies the link between digital media technology and society. His books include *Media and New Capitalism in the Digital Age* (2010), *Internet and Emotions* (2014; co-edited with Tova Benski), and *Reconsidering Value and Labour in the Digital Age* (2015; co-edited with Christian Fuchs).

Routledge Focus on Digital Media and Culture

Algorithms and Subjectivity

The Subversion of Critical Knowledge

Eran Fisher

Routledge
Taylor & Francis Group

LONDON AND NEW YORK

First published 2022
by Routledge
2 Park Square, Milton Park, Abingdon, Oxon OX14 4RN

and by Routledge
605 Third Avenue, New York, NY 10158

*Routledge is an imprint of the Taylor & Francis Group, an
informa business*

British Library Cataloguing-in-Publication Data
A catalogue record for this book is available from the British
Library

Library of Congress Cataloguing-in-Publication Data
A catalog record has been requested for this book

ISBN: 978-1-032-05194-9 (hbk)
ISBN: 978-1-032-05207-6 (pbk)
ISBN: 978-1-003-19656-3 (ebk)

DOI: 10.4324/9781003196563

Typeset in Times New Roman
by MPS Limited, Dehradun

To Uri Ram

Contents

Introduction: Subjectivity redundant

In recent years, almost unnoticeably, algorithms have become close companions. They are now embedded in digital media – from dating sites to navigational applications – hence taking part in almost every realm of life. Algorithms render the plethora of data that individuals create as they use digital media into knowledge about them, primarily in order to provide them with personalized content that matches their personality, taste, and will. This algorithmic knowledge, in turn, shapes how we experience the digital environment, how we see the world, and how we think about ourselves. At the heart of this book is a dual argument. First, the book argues that algorithms create a new way of knowing, which, in turn, changes our fundamental sense of self and our concept of subjectivity. The book analyzes algorithms as epistemic devices, geared toward creating knowledge, which informs users' decisions, preferences, tastes, and actions, and changes the very sense of who they are. Second, in doing so, algorithms subvert a key tenet of modern subjectivity: the participation of the self in creating knowledge about the self, its capacity for mobilizing self-reflection and critical knowledge in order to expand its realm of freedom.

With modernity and with the Enlightenment, the constitution of subjectivity has been entwined with knowledge. Subjectivity – understood here as a quasi-transcendental (both utopian and actual) realm of individual freedom and authenticity – could only be achieved by a self that gained critical knowledge about the world, about social reality, and most prominently about itself. As scientific and humanist knowledge had increasingly replaced theological cosmologies, individuals increasingly participated in creating that knowledge. And as they created knowledge about themselves, they also constituted their subjectivity as a realm of freedom, as that realm that allows individuals to articulate "this is what *I* think/like/want".

DOI: 10.4324/9781003196563-1

Algorithmic decision-making devices have changed not only how knowledge is created but, more radically, what constitutes valid knowledge. In the past two decades, with the increasingly ubiquitous use of digital media, algorithms have become ever more prominent in creating knowledge about what people think/like/want by analyzing their digital data footprint. This knowledge is integrated into everyday life and informs peoples' perceptions and actions. As a result, algorithms increasingly sidestep the role of subjectivity in the formation of that knowledge, undercutting the humanist project of a self able to form and articulate what it thinks/likes/wants. Algorithmic knowledge undermines a key process in the constructing of subjectivity: self-reflection, or the active participation of the self in creating knowledge about the self. It is the self gaining knowledge about the self, which makes such knowledge worthy of the term critical (*á la* Kant), because it is knowledge geared toward widening the scope of its freedom. By bypassing the participation of the self in the creation of knowledge, algorithms also change contemporary subjectivity. The empirical and theoretical inquiry of that transformation is the purpose of this book. The book analyzes the unique qualities of algorithmic knowledge and underlines their divergence from critical knowledge, which emerged during modernity.

The description and analysis of the role that algorithms take in the constitution of knowledge serve as foundations for another discussion that this book offers: the political implications of the rise of algorithms and their subversion of subjectivity. Critical knowledge and the emergence of a fully-fledged subjectivity were a political project; a fantasy – perhaps *the* fantasy – of the Enlightenment. While it could have never been fully realized, striving toward this utopian horizon had deep social, cultural, and political ramifications. Since critical knowledge involved a subject who knows itself, subjectivity also assumed – as well as gave legitimacy to – political agency. And so it is this "mature" human-being, in Kant's famous words, that could exercise her free will. Subjectivity (however imagined and presumed) underlines virtually all central institutions of modernity. In the political realm, it is a subject who is assumed to cast its vote; in the free market, it is a subject who could choose which products to purchase and which contracts to sign; in the realm of science and letters, it is a subject who could form true knowledge and exercise an authentic judgment of taste; in the legal realm, it is a subject who is held liable for its actions; even in the private sphere of marital relations, it is a subject who could choose who to marry or, indeed, whether not to marry at all. Subjectivity and the critical knowledge it assumes are at the foundations of our most cherished modern institutions. And it is this concept of subjectivity, which is now undermined by algorithms.

This book, then, offers an account of how algorithms operate in society as knowledge devices: the new kind of knowledge that algorithms create, the social understanding of this knowledge, and its social ramifications. At the heart of this account is a key question concerning the link between algorithms, knowledge, and self: What are the implications of knowledge without subjects, knowledge, which leaves subjectivity redundant? All five chapters seek to explore the new form of knowledge about the world propagated by algorithmic systems. Spanning different social spheres and case-studies, these chapters ponder most fundamentally about how algorithmic knowledge concerning objects, practices, and ideas in our social world transforms their meaning, significance, and ramifications.

Chapter 1 ties together the question of knowledge and subjectivity, offering a framework for critically engaging with the empirical analysis presented in the following four chapters. It proposes to think about algorithms as *epistemic devices*, socio-technical assemblages engaged in the creation of knowledge about the world and about the self. Given the increasing centrality of knowledge in the conduct of society in the past few centuries, and the ubiquitous presence of algorithmic devices in everyday life, we must ask what is the nature of this new epistemology, and what its social ramifications are. Using Habermas' theory of knowledge, the chapter asserts the inherent inability of algorithms to underlie critical knowledge. It introduces a two-legged argument, which runs throughout the book. The one considers subjective and inter-subjective processes of self-reflection and communication as crucial to the creation of critical knowledge. The other sees algorithmic knowledge as striving to create knowledge about the world and the self by bypassing these subjective and inter-subjective facets. Algorithms strive to create positivist and behavioral knowledge, which does away with self-reflective and communicative facets of knowledge. This, I argue, cannot be understood merely in technical and mathematical terms but should be seen as overturning the modernist teleological utopia, which strove to give the subject a constitutive role in the creation of knowledge, and, in turn, give knowledge a constitutive role in the construction of subjectivity.

The following chapters present four empirical case-studies of algorithmic knowledge. Chapter 2 helps lay bare key assumptions of algorithmic knowledge by asking how algorithms "see" human beings, or more precisely, what the ontological assumptions about humans, which underlie the algorithmic gaze are? It does that by historicizing how media outlets identified and characterized their audience. The media has always been interested in knowing their audience in order to increase the efficacy and reception of their messages. During the era of mass media, "seeing" the audience was based on a scientific episteme combining social theory and empirical research. With digital media, this process has been supercharged by data and algorithms, presumably enabling to know the audience with ever-more detail and accuracy. Our argument is that behind those different technical methods of the mass media and digital media lay also different ontological assumptions about what individuals are. In the mass media, the individual was conceived as having a more or less stable identity, which, in turn, was part of a social group. In digital media, there is a breakup of the conception of the individual as coherent and social; individuals are seen as defined by their ever-changing behavior and are grouped with others who show similar data-patterns. These different ontologies of the individual have deep political ramifications. In the era of mass media, categorization into a social group was of political significance – whether one agrees with that identification on not – because it was done in natural language and it was public. One could argue about the validity of catering for middle-class housewives with particular media content, but at least there was a public concept of a self, on which debates could ensue. Because the categories into which digital media put individuals in order to personalize media content cannot be explained in natural language, and do not involve the self-reflection of the individual, the social, and political meaning of these categories is lost.

Chapter 3 delves deeper into the question of the algorithmic imaginary. It assumes that how we think and talk about algorithms has ramifications to how technology is developed and used. This is doubly true if this imaginary is created and propagated by companies and professionals who play a central role in shaping our digital environment. The chapter offers a reading of *The Selfish Ledger*, a short video created by Google, as a vignette through which to contemplate the ontology of knowledge in the age of big data, algorithms, and artificial intelligence. With the metaphor of the selfish ledger, the video articulates a new relationship between self, knowledge, and media. To unravel the assumptions underpinning this relationship, the chapter recalls the original referent of the ledger as a fundamental logistic media in the development of accounting (i.e., double-entry bookkeeping), and compares it to another pivotal media of modernity, developed around the same time: the personal diary. Both the ledger and the diary shared the assumption that monitoring and registering data in real-time and analyzing them over time yields new knowledge, which is otherwise inaccessible. This was fundamental to the new project of reflexivity dictated by Protestantism. Where they differ radically is the role of the subject in the creation of that knowledge. While the diary assumed the constructive participation of subjectivity in the creation of knowledge, the ledger strove to create knowledge, which bypasses subjectivity. This historical detour helps clarify the utopian horizons represented by *The Selfish Ledger*: a digital media capable of creating knowledge without reflexivity, reason, and critique, that is, without subjectivity. My analytical move here is to situate Google's imaginary – of each user having a ledger containing all her data, which, in turn, directs her actions – within the history of media set on creating new types of knowledge, and their link with a particular subjectivity.

The reality behind the fantasy of algorithmic knowledge, articulated in Google's video, can be glimpsed through two case-studies, explored in the next two chapters. Chapter 4 focuses on recommendation engines and seeks to clarify how the very notion of culture changes as algorithms take part in helping individuals make their own judgments of taste. Rather than seeing them merely as technical and mathematical devices, this chapter argues that underlying recommendation engines are ontological assumptions about culture and aesthetic judgment. The chapter lays bare two key assumptions by analyzing public discussions about Amazon's and Netflix's recommendation engines: that aesthetic judgment is *objective*, that is, that judgments can be obtained by analyzing observable online behavior without tapping subjective contemplation; and that it is *individualistic*, that is, that one's relationship with culture is dyadic, and geared toward satisfying personal needs, without reference to inter-subjective deliberation. The chapter historicizes and relativizes this particular model of aesthetic judgment by comparing it to the modernist model, as formulated in Arendt's reading of Kant. For Arendt, aesthetic judgments are neither objective nor individualistic; instead, they involve subjectivity and inter-subjectivity, since they lay not in the realm of truth, and entails communal, communicative process. By unpacking the assumptions behind recommendation engines, the chapter deepens our understanding of how algorithms are undercutting human agency.

Lastly, Chapter 5 details another concrete example of how algorithms are already becoming independent agents in the formation of our social and political life. It examines the new type of knowledge that algorithms create about space – algorithmic spatiality – and how this knowledge participates in the production of space. Focusing on the navigation giant Waze, it asks how this new socio-technical actor legitimates the knowledge it creates, that is, how Waze asserts its 'right to the city'. The chapter analyzes the clash between Waze and local residents over the applications' common practice of diverting large volumes of traffic through side-roads, quiet neighborhoods and serene villages. These clashes – by legal, political, and discursive means – won media converge across the world, and form the corpus of the research. The chapter shows how along established forms of knowledge, which underlie different actors' right to the city – expert knowledge, democratic knowledge, market knowledge, and local/situated knowledge – emerges a new kind of knowledge, backed by big data and algorithms and managed by a quasi-monopolistic platform, which claims a legitimate right to the production of space. Traditionally a right upheld by underprivileged groups and individuals, the right to the city is currently upheld by a socio-technical assemblage, undercutting the very notion of "rights" as linked with subjectivity.

Put together, the chapters of the book try to answer two key questions about algorithms. First, what is the kind of knowledge that algorithms produce? How do gigantic piles of data and meta-data, produced predominantly by users on digital platforms, get translated into knowledge about them? How is this algorithmic epistemology different from previous epistemologies? And second, how does algorithmic knowledge change our culture, politics, and society? And perhaps more radically, how does it change our very notion of what society, culture, and politics are, and what it means to be human?

My answer tries to go beyond the old adage of "machines overcoming humans" and suggest a more nuanced version. Algorithmic knowledge, I suggest, strive to compete with, supersede, and indeed bypass, a *particular* aspect of being human – subjectivity. It tries to imagine and construct knowledge, which does away with self-reflection, judgment, communicative rationality, and critical knowledge. It is not human-beings as such that are being relegated by algorithmic decision-making; more specifically, what is being relegated is their critical faculties. This trend, I further argue, bares political significance that we must grapple with.

* * *

This book culminates a research project that began in 2015. Research for this book has been largely funded by a generous support from the Israeli Science Foundations (No. 696/16) entitled "How algorithms see their audience?" Parts of this book have appeared in earlier versions in *Media, Culture, and Society*, *Continuum: Journal of Media and*

Cultural Studies, and *Cultural Studies*. My thanks to the respective journals for granting permission to reprint them here.

Many people and institutions provided me aid, advice, and camaraderie during this time, and it is a pleasure to thank them. Thank you to my colleagues and friends at my home institution, the Open University of Israel. Special thanks go to Anat Ben David and Zeev Rosenhek, my close scholastic companions. Thank you also to Ishay Landa for reading an earlier version of Chapter 3 and providing invaluable suggestions. Thank you to Laurence Barry, Norma Musih, and Hemi Ramiel, post-doctorate colleagues with whom I collaborated on parts of this project. Their intelligence, knowledge, and good companionship have had an invaluable contribution to my work. Thank you to my colleagues who have took the time to hear or read some of the materials that make up this book. Thanks to The Van Leer Jerusalem Institute for hosting a two-years seminar on Political Theory and Science and Technology Studies (led by Anat Ben David and myself) and especially for its regular participants, who have also commented on parts of this book. Special thanks to Ilan Talmud for his incisive comments on my work. Thank you to Bjarki Valtisson and Ivo Furman with whom I shared a grant from The Danish Agency for Science and Higher Education, as well as many ideas and laughs. Thank you to The Institute for Cultural Inquiry (ICON) & The Department of Media and Cultural Studies at Utrecht University in the Netherland, where I've spent 2021 as a guest researcher. I'd like to thank particularly José van Dijck and Frank Kessler for their hospitality and generosity. Lastly I want to thank the people in my inner circle who make life valuable. Thank you to Tamar Assal, Shira Dvir-Gvirsman, Efrat Eizenberg, Nitzan Famini, the Feinsteins (Orit, Lior, Yuval, Nadav, and Noa), Angela Fette, Ben Fisher, Shenja van der Graaf, Dana Kaplan, Yael Lev, Gal Levy, Yoav Mehozay, Sharon Ringel, and Nava Schreiber. Above all, thank you to my parents, Rina and Carol Fisher, and thank you to my children, Yuli van der Graaf, Ira Paz, and Roya Fisher.

I dedicate this book to Uri Ram, my teacher and mentor. His invaluable contribution to my scholarship has become ever more apparent to me the more my topics diverged from his. His commitment to science as a humanistic project with an emancipatory horizon has been a beacon for me; his friendship has been a precious source of warmth in my life.

1 Can algorithmic knowledge be critical?

Algorithms and you

Imagine turning on your Netflix in the evening to find out it had put on a movie, which fits perfectly with what you'd want to watch. It not only hits the right keys of your taste in movies and your recent and changing aesthetic interests but also it seems to take into account the specificities of your daily happenstances and mood. But you are not particularly surprised. You remember Netflix's chief executive officer (CEO) Reed Hastings' pronouncement that "One day we hope to get so good at suggestions that we're able to show you exactly the right film or TV show for your mood when you turn on Netflix" (*Economist,* 2017). You are also aware of the efforts and technological agility involved in reaching such a phenomenal knowledge of your taste, wants, and mood. Netflix, you know, monitors the data traces you leave on its platform. Maybe, you'd assume, Netflix complements it with data it gathers from other digital media, such as social networking sites. This big data set – about you, as well as about all its more than 200 million users – is crunched in real-time by algorithms, which are able to know not only who you are, what your taste in movies is, and so forth, but also discern very accurately your wishes, desires, and needs per a particular moment. Maybe, their choice of a Hollywood romantic comedy from the 1950s would have been different lest you were sitting there with your lover on a Thursday night. Who knows? But should you even care? After all, the match is perfect. As perfect, in fact, as the match of a dating site, which introduced you to your lover but a month ago. There, too, you'd assume a plethora of data (some of which provided by you) has been processed algorithmically to salvage you from your own failed attempts to find a suitable partner.

DOI: 10.4324/9781003196563-2

Indeed, encapsulated in the digital devices we use – or better say, the digital environment we inhabit – is a promise to better the human condition and expand its convenience and contentment. With modernity, technology has come to play not only an instrumental role – making things and processes more efficient, quick, or at all possible – but an ideological role as well. Whether Left-leaning and progressive or Right-leaning and conservative, across virtually the whole spectrum of modern politics technology has come to be seen as means for political ends and as their guarantor. Technology promised to allow the fulfillment of the ideals of modernity and the Enlightenment, immortalized by the French Revolution's adage: *Liberté, Egalité, Fraternité.* Different political orientations defined these ideals differently, but they shared an underlying ideology, which sought to mobilize technology – that is, applicable, scientific know-how – in order to secure their materialization.

In *Media and New Capitalism in the Digital Age* (Fisher, 2010), I have shown the transformation in technology-*cum*-ideology with the rise of digital media. Whereas mechanical and centralized production technology, which dominated the industrial age, was seen as demanding a Fordist, Keynasian, social-democratic contract (as well as securing such a contract), digital, distributed, information and communication technology of the early digital age was seen as demanding and securing a post-Fordist, neoliberal social contract. At the heart of this new technology ideology were keywords such as distributed networks, de-hierarchization, flexibility, and adaptability, all of which allowed, demanded, and secured a neoliberal order. The current book, *Algorithm and Subjectivity*, continues this theoretical thread but turns its gaze to a more specific characteristic of digital technology: its ability – indeed propensity – to algorithmically render user-generated data into usable information and knowledge. I term this algorithmic knowledge, or algorithms, as a shorthand.

By algorithms, I mean a socio-technical assemblage geared toward rendering data into information and knowledge. This understanding is both wider and narrower than what the common use of the term suggests. Wider, because algorithms refer in this book not merely to lines of code, which render input into output in order to receive a desired outcome. Rather, by algorithms, I mean a whole socio-technical assemblage of people, technologies, practices, sites, and knowledges. This includes the incessant production and accumulation of big data in digital sites, predisposed to collect user-generated data (platforms); the construction of technological tools, which make sense of this data, turning massive amounts of personal data into knowledge (algorithms, machine learning, neural networks, artificial intelligence), bodies of knowledge concerning these practices (e.g., data science), professionals, and executives. But my use of the term is also narrower in that it refers to the use of algorithms in digital media, specifically, algorithms integrated into online decision-making devices, or *interface algorithms* (more on that below).

Technologically, the promise encapsulated in algorithms is that by letting algorithms sip through the plethora of data, inadvertently created by users, they could determine who the users are and what their needs and wants are. But this technical promise to automate knowledge about the self – as part of a larger project to create algorithmic knowledge about the world – goes much deeper. It is ultimately an ideological promise to make us freer, more emancipated human beings.

But what does this human freedom entail in the context of digital media? Is it a promise to free us from the *Burden of Choice* (Cohn, 2019)? Since the emergence of the internet, our choices have been expanded exponentially, with access to a virtually endless array of information, cultural artifacts, products, and people. To paraphrase Chris Anderson, the monster of choice that had awaited us in the supermarket aisles has now grown a long tail over the internet (Anderson, 2006). With this promise for an endless supply of information goods we face an abyss: How are we to choose? With algorithmic devices, digital platforms created the solution to a problem of their own making. Algorithmic curation, recommendation engines, and the social graph were all new means put in place presumably to allow a happier marriage between users on the one hand and information resources – cultural, political, economic, and social – and the world at large, on the other hand.

But this seemingly technical solution underscores a promise, which is at an even more fundamental level: to bring individual subjects back into the scene, to facilitate the constitution of each of us in the wired world as a unique individual. Perhaps no other keyword in the universe of digital media reflects this promise more than *personalization*. After a century of mass-communication – which grew out of mass society and mass culture, as well as pampered it – digital media is able to offer each and every member of the masses his and her own bouquet of mediated artifacts, be them movies, consumer products and service, advertisement, or otherwise.

Know thyself

Most fundamentally, then, algorithms promise to expand the realm of personal freedom by offering a truer, richer, more precise knowledge about the world and about our self. The idea that deepening one's knowledge also expands one's freedom was born with the Enlightenment. It is a specific articulation of the more general promise of knowledge (i.e., science and technology) to better the human condition. Modernity and the Enlightenment offered new forms of knowledge about the world and the self. Knowledge that involves self-reflection and expands self-understanding by engaging the self in deciphering the self. This new encounter of the self by the self, stimulated by self-reflection, is what I call here subjectivity. Subjectivity has always been a promise. A promise born in the Enlightenment to expand the realm of freedom from natural instincts and impulses, as well as from human-made coercive and oppressive social relations. Arguably, this promise has never been – and could never be – materialized to the fullest. But it nevertheless offered a horizon for what human freedom might mean. Subjectivity was not seen as ontological, a reality to be discovered, but rather as a project worthy of being achieved.

Digital media now offer a new model of knowledge, based on the algorithmic processing of big data gathered mostly by using this very media. If, for the past few centuries, self-reflection, a self, which knows itself, has been the cornerstone of subjectivity, which was, in turn, a precondition for freedom, my question in this book is: What kind of freedom underlies algorithmic knowledge? If *Know Thyself* was a route for a more emancipated subjectivity, what kind of freedom is promised by algorithms mediating for us knowledge about the world and about our self?

I argue that compared with this ideal of the Enlightenment, algorithms offer a very different conception of knowledge and subjectivity, a different imaginary (Bucher, 2016). Algorithms, I show throughout the book, offer not merely a new method and methodology to answer questions. Rather, they offer a new epistemology, which redefines what questions can be asked, and what it means to know. The Enlightenment's ideal of knowledge, particularly knowledge about the self, was inherently critical. This thrust is epitomized by Kant's three volumes of *Critique*. In the *Critique of Pure Reason*, Kant (1999) sought to lay bare the structural epistemological conditions and transcendental assumptions, which frame common and accessible conceptions of empirical knowledge, which is deemed true and valid. In his *Critique of Practical Knowledge* (2015), he sought to define the limiting conditions and assumptions, which dictate our behavior and to excavate the conditions of autonomy and morality. Kant, thus, signifies the high-point of Enlightenment, which enables both the development of objective knowledge and all the while guarantees subjective emancipation.

In contract, I argue, the algorithmic model of knowledge is one-sidedly based on positivist assumptions, which impels it to exclude subjectivity from knowledge about the self. Rather than promoting an interpretive, hermeneutic, and reflective approach to the self, it suggests to exclude subjectivity from such an endeavor. Instead, it suggests that we will be most authentic to our true self if we let algorithms tell us who we are. With the advent of algorithms and the interweaving of our existence with digital devices, which, in turn, gives us access to huge quantities of data, indicating actual behavior, the argument goes, we are in a unique epistemic position to know our selves better than ever before.

The purpose of this book is not to assess whether such a task has been achieved, nor whether it is at all possible. In fact, the question I pose makes such judgment irrelevant. I ponder, instead, what can be defined as *The Political Unconscious* of algorithmic culture (Jameson, 1981). With that task in mind, I ask: To the extent that the promise of algorithms – predicting which word we'd want to type next, what movie we'd like to watch, who we'd be interested to date, and so forth – materializes, what are the horizons of this promise in terms of our conception of subjectivity? My short answer is that it is a form of knowledge about the self, which ultimately excludes the self from the process of learning and knowing about the self, that is, excludes self-reflection, and in so doing subverts the Enlightenment project of subjectivity. To a large extent, an algorithmic social order requires no subjects at all, but rather seeks to turn them into objects. Karl Marx, a philosopher who supplemented Kant's line of *Critiques* with his *Critique of Political Economy* (the subtitle of his magnum opus *The Capital*) (1992), had already warned against the subsumption of the (working) subject under the (produced) object, a process that he termed the fetishism of commodities.

It is tempting, at this point, to read my argument as a reiteration of the old adage of humans versus machines, or rather, technology taking over humans (Winner, 1977). This, too – that is, not only technology as promise but also as peril – has been a staple of modernist though, epitomized by the likes of Heidegger (1977a) and Ellul (1964) (see also: Borgmann, 1999; Postman, 1993). However, my argument concerning algorithms strives to diverge from such analyses, which Robins and Webster aptly describe as "technologistic" (Robins & Webster, 1999). While my analysis seems to reach similar gloomy conclusions concerning algorithms, it finds the culprit not in "technology" as such, but in a specific constellation thereof. The threat of algorithms to subjectivity does not stem from the mere fact that knowledge about the self is mediated. In fact, there has been a long history of media devices that have been helpful in creating knowledge about the self *with* the self, thus contributed to self-reflection and helped expand the realm of subjectivity (more on that in chapter 3). What these epistemic media share is the engagement of the self in the creation of new knowledge, which has led, either intentionally or as a side-effect, to the opening up of a new space of self reflection. Algorithmic devices, in contrast, as a very particular type of epistemic media, exclude the self from knowledge about the self, or rather reproduce the self as a media-made artifact.

Algorithmic knowledge and human interests

There is no doubt that, given the right resources, algorithms are able to create knowledge. The question is what is that knowledge and what is its truth value, that is, under which assumptions is this knowledge valid. From the perspective of the social sciences, questioning algorithmic knowledge has focused predominantly on the nature of that knowledge, how it differs from other epistemologies, and what are the ramifications of increasingly integrating algorithmic knowledge into the social fabric. Algorithmic knowledge has indeed been criticized for its biases (Crawford, 2016; Gillespie, 2012a, 2012b; Mayer-Schönber & Cukier, 2013). Such biases may have detrimental social consequences from distorting our image of the world to racial discriminating (Ferguson, 2017; Gillespie, 2016; Mehozay & Fisher, 2018; Tufekci, 2019). What is more, their opacity makes public audit and critique of them virtually impossible (Kim, 2017; Mittelstadt, 2016; Pasquale, 2015b; Soll, 2014). Algorithmic knowledge has also been criticized for creating and perpetuating a feedback loop for users, enclosing them in a *Filter Bubble* (Pariser, 2012; Turow, 2011). And given their underlying political economy and their reliance on personal data, algo-rithms have also been criticized for inherently undermining privacy (Dijck van, 2014; Grosser, 2017; Hildebrandt, 2019; Kennedy & Moss, 2015), and for exploiting audience labor (Andrejevic, 2012; Bilic, 2016; Fisher & Fuchs, 2015; Fuchs, 2011b). All these point to algorithms as constituting a new regime of knowledge, which has a huge impact on contemporary life, yet remains largely unknown, unregulated, and outside of the realm of democratic politics (Feenberg, 1991).

Yet there is another type of critique of algorithms. As research concerning various social fields has shown, algorithmic knowledge does not merely automate the process of knowledge creation but changes the very ontology of that knowledge. For example, algorithms implemented in the cultural context, such as recommendation engines, also change the very meaning of culture, as well as cultural practices (Anderson, 2013; Bail, 2014; Gillespie, 2016; Hallinan & Striphas, 2014; Striphas, 2015). Last but not least, the self – the characteristics and qualities of which so much of algorithmic knowledge in digital media is oriented to decipher – is not merely gauged and monitored by algorithms, but is also altered (Barry & Fisher, 2019; Cheney-Lippold, 2011; Fisher & Mehozay, 2019; Pasquale, 2015a).

* * *

This book joins this last line of critique, which sees algorithmic knowledge as constituting a new epistemology, a new way of knowing. My understanding of knowledge – underlying the various engagements with algorithms in the following chapters – and its relation to subjectivity, draws predominantly on Jürgen Habermas's theory of knowledge, in particular his book *Knowledge and Human Interests* (Habermas, 1972). Before discussing his theory it's worthwhile recalling the state of knowledge – both in society and in social theory – which has prompted Habermas to offer his interjection.

Habermas reacted to what he considered to be a dual attack on knowledge. At the time of the book's publication in 1968, knowledge was becoming an important axis in sociological theory and would remain dominant for a few decades to come, as revealed by keywords such as post-industrial society, information society, knowledge society, network society, knowing capitalism (Castells, 2010; Mattelart, 2003; Stehr, 2001; Thrift, 2005; Webster, 2002). Knowledge was beginning to be understood as located at the center of a radical shift in the social structure of Western societies. This was a view shared by schools of diverse paradigmatic approaches and political affinities. The most notable sociologist to theorize the emerging centrality of knowledge in determining the social structure was Daniel Bell. A post-industrial society, Bell proposed, where knowledge and information gain an axial role in the organization of society, sees the rise of a rationalized class of professionals, and of a technocratic government, both bent on applying knowledge to solve political problems (Bell, 1999; Touraine, 1971). Such a society is managed more rationally, overcoming the ideological struggles that characterized the industrial society.

Bell's claim for a radical break in the social structure was coupled by post-structuralists' claim for a radical break in social epistemology, brought about by the centrality of knowledge – as well as infortmation, symbols, data, myths, narratives, and so forth – in society. Post-structuralism undermined the hitherto *sine qua non* of knowledge, its representationality: the capacity of knowledge (in principle if not in reality) to correspond with reality. In the formulation of Lyotard (1984) and Baudrillard (1981), knowledge – particaulrly due to the introduction of information technology – was becoming a central axis of the social to a point of overwhelming the reality it is supposed to reflect. Joining Foucault (1994) and Derrida (1974), knowledge was now seen as explained better by reference to power relations than by appeals to reason and truth, thus losing its analytical distinction from power.

Both positions, then, undermined the *critical* potential of knowledge, its potential to transform society. Post-industrialism de-politicized knowledge, imagining it as a monolithic social endeavor, which makes politics redundant. Post-structuralism politicized knowledge to such a degree that it invalidated its autonomy from power. In both formulations, knowledge has become a force for conserving and stabilizing power relations. Or put somewhat differently, whereas Bell and other structuralists conceived knowledge as allowing the rationalization of society by making ideologies irrelevant, post-structuralists expressed deep disbelief in knowledge as a rationalizing agent, insisting on its interlacing with power. As both accounts also acknowledged the centrality of knowledge in contemporary society, this was not a happy predicament to a critical social theorist, such as Habermas, whose vista has been the resurrection of the enlightened subject and rational inter-subjective communication.

Habermas sought to offer a *critical* theory of knowledge, which, at one and the same time, upholds knowledge as a vehicle for rationalization *and* accounts for its ability to transform reality toward a horizon of emancipation. How can knowledge be committed to both (scientific) "truth" and (political) "enamcipation"? Habermas' solution is to suggest that knowledge is inextricably linked with human interests. In other words, *all* knowledge is political; it inevitably operates within the confines of human ends. The choice of the term "interests" in the title of Habermas' book is illuminating and makes for three different readings. "Interest" can refer to a sense of intellectual curiosity and a drive to understand reality; "knowledge for the sake of knowledge" (Habermas, 1972, p. 314). Such a reading would suggest that Habermas is concerned with what individuals and societies are interested in. "Interest" can also refer to having a stake at an issue, to standing to gain or lose something. That would suggest that the book title refers to what individuals and societies have a stake in. Finally, the title could also mean both and suggest, as I think Habermas does, that what humans are curious about is inextricably linked with what serves their interests. It suggests that we cannot dissociate the history of knowledge from the political contours within which humans seek this knowledge. To use a later formulation, Habermas suggests that rather than denying, condemning, or duly accepting the knowledge/power nexus, it should instead be examined and theorized. And that's what Habermas sets out to do.

Habermas identifies three types of "knowledge interests" – that is, motivations to gain knowledge – each stemming from human existence and having come to be articulated in three types of scientific or scholarly inquiry. (That also means that knowledge is not monolithic, as Bell suggests, but multiple). The first is a "technical interest", our species' survivalist interest in controlling and predicting our natural environment. This interest has given rise to the "empirical-analytic" sciences, mostly the natural sciences, but also streams in the social sciences that have been modeled after the natural sciences. This knowledge approaches nature and society as objects, which are governed by predictable regularities, and which can therefore be discovered by controlled methodologies (e.g., experiments), articulated into law-like theories, and even manipulated through intervention.

Second is a "practical interest", which involves the attempt to secure and expand the possibilities for mutual understanding in the conduct of life. This interest gives rise to the "cultural-hermeneutic" sciences, a type of knowledge that presupposes and articulates modes of personal and inter-personal understanding, which are oriented toward action. Such understanding is not "scientific", or "objective" in the common sense, but concerned with the lifeworld and is expressed in the grammar of ordinary language. It is exercised in realms of knowledge such as history, anthropology, and parts of sociology and communication studies. Both the empirical-analytic sciences and the cultural-hermeneutic sciences are academically established and constitute a hegemony of knowledge.

But Habermas wishes to go beyond this hegemony by pointing to another deep-rooted human interest, which has given rise to another form of knowledge. This is the "emancipatory interest" of reason, an interest in overcoming (externally-imposed) dogmatism, (internally induced) compulsion, and (inter-personal and social) domination. The emancipatory interest gives rise to critical knowledge. Critical knowledge has a few defining features that Habermas would go on to examine in later works, most notably in *The Theory of Communicative Action* (Habermas, 1985). Particularly crucial to our discussion here is self-reflection, i.e., the central role of the knower in the creation of knowledge. Creating critical knowledge about human-beings (as social, anthropological, or psychological begins) is a *praxis*, which requires the participation of the objects of that particular kind of knowledge, i.e., human-beings. Critical, emancipatory knowledge involves, therefore, subjectivity as both a precondition and an end-product. Critical knowledge can only emerge with the involvement of subjectivity; subjectivity, in turn, can only emerge with critical knowledge.

With the notion of critical knowledge, Habermas sought to offer a category of knowledge, which accounts not merely for reality, but also for the conditions under which this reality comes about and is made possible. Such knowledge can then serve to inform actions needed in order to change these conditions. It is therefore at once both objective and positivist (appealing to truth) and subjective and constructivist (appealing to power). As McCarthy notes in the introduction to Habermas's *On the Logic of the Social Sciences*, Habermas "finds that the attempt to conceive of the social system as a functional complex of institutions in which cultural patterns are made normatively binding for action" – a description corresponding more or less to Talcott Parsons' by-then hegemonic social theory (1968) – "does furnish us with important tools for analyzing objective interconnections of action; but it suffers from a short-circuiting of the hermeneutic and critical dimensions of social analysis" (McCarthy, 1988, p. viii). In other words, such theory excludes the communicative, subjective and inter-subjective dimensions of society, where actors reflect upon their actions, and are able to critique them.

It is worthwhile noting that Habermas does not critique positivism *per se*, as a mode of scientific inquiry. Rather, his critique is more nuanced: he rejects positivism's claim to represent the only form of valid knowledge within the scientific community, and more acutely, its application to concerns, which require critical knowledge. There are obviously concerns which require a strategically oriented action, demanding instrumental reason and constituting subject-object relations) (e.g., ensuring a given growth rate of the national economy). But such type of action, Habermas insists, must not colonize concerns, which require communicative action, demanding communicative reason and constituting subject–subject relations (e.g., questioning whether economic growth is desirable, or even what constitutes "growth" in the first place).

With critical knowledge, Habermas calls for the uncovering of that which not-yet-is, and which may-never-be unless we notice it and made knowledge about it explicit. This is the Schrödinger's cat of the social, the political, and the cultural. And whether we find out the cat is dead or alive depends on our epistemology, that is, our understanding of what knowing is:

> In the framework of action theory [*à la* Parsons], motives for action are harmonized with institutional values.... We may assume, however, that repressed needs which are not absorbed into social roles, transformed into motivations, and sanctioned, nevertheless have their interpretations. (McCarthy, 1988, p. viii)

One of these "cats", which can hardly be noticed by sociological action theory, is subjectivity, an elusive construct, which is always in the making and which only through self-reflection can gain access to critical knowledge, which, in turn, will realize its emancipatory interests. The moment we start to ask ourselves about our self, we also construct it and change it.

The algorithmic knowledge that makes the focus of this book is not primarily scientific.[1] But as the production of "epistemic cultures" (Knorr Cetina, 1999) and of epistemic devices (Mackenzie, 2005) are no longer the hegemony of academia and books, but of the digital industry and software, we must take account of the kind of knowledge that algorithms create and how this knowledge shapes human understanding of the world and of itself. Similarly to the different theoretical schools that have come to grips with the centrality of knowledge in the reformation of the social structure since the 1950s, we must now acknowledge a new phase in that historical era. In this new phase of algorithmic devices, technology automates not merely human physical force, dexterity, and cognitive skills, but also tenets of our subjectivity and inter-subjectivity; automation which makes them redundant in the conduct of human life.

The performance of algorithmic knowledge

My choice of algorithms as an axial concept seeks to highlight the epistemic character of our contemporary techno-social order, i.e., their orientation toward rendering data into knowledge. The choice of algorithms as a vignette through which to examine our digital civilization stems not only from the increasingly central role that knowledge has come to play in society but also from its ubiquity and banality, that is, its integration into literally every sphere of life. This makes the knowledge that algorithms create about the world not merely Platonic and descriptive but also performative. In fact, as users, we encounter not so much the knowledge that digital media create about us, but the effects of this knowledge, such as our newsfeed on social media sites or a book recommendation.

In professional and public lingo algorithms are often described as predominantly predictive devices. Digital platforms seek to know their users' tastes and wants in order to be able to make personalizations. But the political economy behind digital platforms suggests that their goal is not to predict behavior as much as it is to control it. Control means different things in different contexts. In the case of Amazon, for example, the goal of prediction is to make users purchase a product they would *not* have purchased otherwise. To put it boldly, Amazon's algorithms are oriented to predict not what users want, but what they don't want.

Because algorithms are future-oriented, because they seek to predict behavior and control it, they also seek to ascertain a particular type of subjectivity, which is predictable. To the extent that subjectivity is an important source for self-conduct, and self-reflection may change behavior, algorithmic prediction would be much less successful. For algorithms to deliver on their promise to know who we are and what we want, they must also assume a dormant subjectivity, a subject that is really more of an object (Fisher & Mehozay, 2019). Algorithmic knowledge, then, is performative in the deepest sense: it attempts to imagine and mold a human-being that is completely transparent and predictable. It describes only that which it can control. If algorithmic machines are becoming – or imagined to become – more accurate, it is not merely because of technological advances. Rather, it must also be attributed to the part they play in helping create a self, which trusts algorithms and the knowledge they reveal about it, and which, in turn, sedates mechanism of self-reflection and self-knowledge, precisely these faculties of the self that are potentially opening up a realm of freedom and make humans unpredictable and able to change.

And here, the deep political ramifications of the algorithmic subversion of subjectivity become more evident. Underlying the creation of critical knowledge about our reality is a human *interest* in transforming that reality, and a human *involvement* in creating this knowledge. If subjectivity is a realm of emancipation through critical knowledge, it is at the same time a precondition for critical knowledge to come about. Such is the case, for instance, in Hegelian-Marxist theory, which makes a distinction between class-in-itself (i.e., an objective reality of historical materialism) and class-for-itself, which involves a subjectivity, transformed by that objective knowledge, and which, at the same time, constitutes the agent of social transformation. Such is also the case with Freudian psychoanalysis where self-knowledge is key to self-transformation. Psychoanalysis proposes that one's behavior, thoughts, and desires do not reveal the full scope of who one is; they are certainly not equal with one's true self. As much as our behavior reveals who we are, it also tells us what hinders us from being free, because it stems also from these hindrances. Enlightenment, in the sense of self-reflection, is supposed to make the self aware of these hindrances to freedom, with the hope of transforming the conditions for their persistence.

A self, structured within the contours of an algorithmic environment, is imagined in a radically different way from the self that was imagined during modernity, and has reached its most eloquent articulation with the Enlightenment idea of subjectivity. It is hard to imagine the rise to dominance of algorithmic knowledge in a world populated by human-beings keen on self-reflection in order to expand their subjectivity. Under such circumstances algorithms would not work well. Firstly, they would be rejected as unacceptable avenues for achieving freedom because they exclude the subject. And secondly, under conditions of reflexivity, algorithms would have a harder time predicting wants. By claiming this, I do not mean to play a *what-if* game with history. Rather, I wish to point out that algorithms assume and imagine a particular type of human-being, with particular horizons of (non)subjectivity and (un)freedom. It is the purpose of this book to register and analyze these assumptions, and ask what conception of subjectivity underlays the algorithmic model of knowledge. Owing to the performativity of algorithms, will they succeed in molding a new kind of person and a new self? As counter forces are also always in play, this is a struggle to be fought rather than to be either already celebrated or decried.

Subjectivity, algorithms, and privacy

The juxtaposition of algorithms and subjectivity sheds a new light on privacy and why we should be worried about its erosion. Subjectivity, this space of self-reflection – of thinking about one's thoughts, evaluating ones' desires and wishes, critically assessing one's tastes, and so forth – we could have also called the private sphere, lest this term was already occupied by a somewhat different meaning. Subjectivity can be thought of as a private sphere, where thoughts, wants, and needs of the self can be reflected upon and evaluated by that very self. It is a space that allows, at the very least, a possibility to question our self. The loss of privacy also entails undermining our ability to develop and maintain that space. It is private not merely in the sense of ownership, that is, that whatever takes place in this sphere is mine (like "private property"), but also in the sense that it is autonomous and distinct from other social spaces, and impenetrable for them (like "private matters"). Just as we think about the public sphere as a space that facilitates communicative action, we could also think about subjectivity as a sphere that facilitates an internal critical dialogue. And just as Habermas described the contraction of the public sphere more than half a century ago (Habermas, 1991), we might describe subjectivity as a private sphere that is now facing an attack by algorithms. Simply put, algorithmic knowledge, and its inherent erosion of privacy, also erodes our ability to protect our subjectivity.

As algorithms seemingly try to gauge what takes place in the private space of subjectivity it also contracts that space; it destroys that which it seeks to capture. The innumerable and varied data points that algorithms are able to gauge presumably serve as proxy for that internal space, and get to the crux of personal wants, desires, and needs. But what algorithms cannot gauge is precisely the critical, reflexive events that take place in that space, which allow a dialogue between, on the one hand, what one thinks and wants, and on the other hand, what one thinks about one's thoughts, and how one wishes to deal with one's wants. This space of reflection, of making the self an interlocutor of the self, is not, as aforementioned, an inherent and given component of our humanity. Instead, it is a historical construction, a project of the Enlightenment, and a utopian ideal at that. By excluding this space from the understanding of the self, algorithmic knowledge also undermines this project.

Interface algorithms

The empirical scope of algorithms presented in this book is quite selective, and stems from my disciplinary embeddedness in media studies. I am interested in what could be termed *interface algorithms*: algorithms embedded in digital platforms, such as online retailing sites, social networking sites, and social media. Interface algorithms are geared predominantly toward rendering users' data into knowledge about them and, in turn, creating a personalized interface for each user. This, as aforementioned, is dependent not merely on getting to know users more intimately and intensely than before, but also *differently*, that is, on redefining what such knowing entails. I draw here on literature which sees algorithms primarily as epistemic devices. As knowledge-making machines, algorithms see reality in a particular way, different from modes of knowing we have become familiar with. They offer what David Beer beautifully termed a *Data Gaze* on reality (Beer, 2019), reconceptualizing and redefining that which they see (Beer, 2009; Kitchin, 2017; Mackenzie & Vurdubakis, 2011). This has been substantiated in recent empirical research in relation to media audience (Fisher & Mehozay, 2019; Hallinan & Striphas, 2014) advertising (Barry & Fisher, 2019; Couldry & Turow, 2014) retailing (Turow, 2011; Turow & Draper, 2014), risk in the context of the criminal justice system (Mehozay & Fisher, 2018), and health in the context of medicine (Ruckenstein & Schüll, 2017; Van Dijck & Poell, 2016), to name a few fields.

But interface algorithms incessantly project the knowledge they create about users back at them; they act as mirrors, reflecting users' self. Users are learning to employ an "algorithmic imagination" (Bucher, 2016) to see the content they are offered as an indication of how they themselves are seen by the algorithms, and to some extent (albeit with potential critical distance) as an algorithmic reflection of their self. For example, the fear of remaining "invisible" to their friends on social networking sites disciplines users into productive participation, and shapes their media practices (Bucher, 2012). Algorithms' inner workings may be opaque but their effects are very present, as Bucher puts it. More broadly, Gillespie argues that "the algorithmic presentation of publics back to themselves shapes a public's sense of itself" (Gillespie, 2012b). And Neyland suggests thinking of interface algorithms as "a configuration through which users and/or clients are modeled and then encouraged to take up various positions in relation to the algorithm at work" (Neyland, 2015, p. 122).

This is not to say that people are duped by algorithms, but that they indeed find themselves in an inferior epistemic position to critique the new kind of knowledge they create. Algorithmic knowledge bares the aura of a superior model for representing reality. Most pertinent to our case, perhaps, is the key promise of algorithms to produce knowledge with no *a priory* conceptions, either normative or theoretical (Mayer-Schönber & Cukier, 2013). According to this increasingly hegemonic ideological discourse (Mager, 2012, 2014), by perusing billions of data-points in search of discovering mathematical patterns, algorithms let data "speak" for themselves, thereby purporting to offer a much more objective mode of knowing, evading biases which stem from human-centric normative predispositions. The fact that the basis for algorithmic knowledge is raw data – a "given", as the etymology of the word suggests, a seemingly unobtrusive reflection of reality – contributes to their flair of objectivity (see Gitelman, 2013 for a critique of that assumption).

Subjectivity redundant

Our subjectivity, then, is under attack by algorithms. Or is it? Skeptics may argue that even if my analysis of the case studies presented in the following chapters is valid, my overarching argument is overstated, since subjectivity should not be seen in the first place as a space of emancipation, but as a disciplinary mechanism of governmentality, shaped in accordance with hegemonic social structures (Foucault, 1977, 2006). In other words, we have never been modern and have never been free; subjectivity is nothing but another form of social control.

This Foucauldian line of thought is important to pursue, and I have indeed implemented this mode of inquiry throughout the book. It opens up another interesting avenue for understanding the algorithmic episteme as algorithmic governance (Birchall, 2016; Danaher et al., 2017; McQuillan, 2015; Rona-Tas, 2020; Sauter, 2013). But this avenue, too, leads to a similar conclusion concerning the redundancy of subjectivity in an algorithmic environment. If governing, or the exertion of power, during modernity demanded the willing cooperation of subjects, then algorithmic governance makes such governmentality redundant. Subjectivity was required to keep particular social structures intact and allow them to mobilize individuals into particular social forms and actions. At the same time, of course, such subjectivity – for example, a neoliberal persona – could also be a site of resistance and change.

Much like Habermas, then, Foucault too sees in subjectivity not merely an effect of power but a space capable of resisting power and opposing it. And not unlike him, he too posits knowledge at the very center of subjectivity. The interests may be different – disciplinary rather than emancipatory – but the mode of operation is self-reflection, the dictum to *know thyself*. In this formulation, too, algorithms can be said to interject and change the subject. They expropriate the conduct of conduct from subjectivity, literally conducting behavior. If subjectivity harbors the commends that tell us how to conduct ourselves, then the introduction of algorithms conducting our conduct, governing it externally, makes subjectivity redundant.

Following this link of inquiry, too, leads us to consider how algorithms undermine subjectivity. Algorithms become a new governing agent, which manages life and populations without the need for subjectivity. Under such conditions, does it make sense to talk about algorithmic subjectivity at all? Should we not, as Rouveroy suggests, think about the effects that algorithms pursue as creating objects rather than subjects? The political ramifications stemming from this line of though are troubling. As Rouvroy and Stiegler (2016) and Hildebrandt (2019) have convincingly argued, algorithms make claims for sovereignty of a new kind, as they are able to take decisions that are almost impossible to audit and critique, because they are opaque, proprietary, and subject to frequent change. It is therefore our task to critique algorithms' participation in social life and their claim for political rights through an interrogation of their political effect.

Knowledge about human beings – that is, the knowledge of the human sciences, which is at the center of most of Foucault's works – changes radically. Social epistemology, as we might call it, shifts from regimes of truth to regimes of anticipation, which are increasingly dependent on predictive algorithms (Mackenzie, 2013; Rona-Tas, 2020). In such regimes, "the sciences of the actual can be abandoned or ignored to be replaced by a knowledge that the truth about the future can be known by way of the speculative forecast" (Adams et al., p. 247, cited in Mackenzie, 2013). Knowledge, in the case of the algorithmic episteme, boils down to the ability to anticipate future trends and patterns. This entails seeing individuals based on the behavioral data they produce (Rouvroy & Stiegler, 2016), bypassing their self-understanding and identifying patterns from which a predictive behavioral analysis can be deduced.

There are many elements that make algorithmic knowledge an unsuitable foundation for critical knowledge, such as refraining from theory, and from an ontological conception of humans (Fisher, 2020). But most fundamental is the attempt of algorithmic knowledge to bypass subjectivity en route to the creation of knowledge. That is, to create knowledge about the self, which does not allow the subject – for lack of ability to use natural language – to audit such knowledge with the aid of reason. That is true to human knowledge in general, but it is doubly true for their knowledge about themselves, as social, anthropological, and psychological beings.

Note

1 It is worthwhile mentioning that the use of algorithms for scientific endeavour is on the rise also in the social sciences and the humanities. See, for example, Alvarez (2016), Dobson (2019), Levenberg et al. (2018), Marres (2017), and Veltri (2019).

References

Alvarez, R. M., ed. (2016). *Computational social science: Discovery and prediction.* Cambridge: Cambridge University Press.

Anderson, C. (2006). *The long tail: Why the future of business is selling less of more.* New York: Hyperion.

Anderson, C. W. (2013). Towards a sociology of computational and algorithmic journalism. *New Media and Society*, *15*(7), 1005–1021.

Andrejevic, M. (2012). Exploitation in the data mine. In C. Fuchs, K. Boersma, A. Albrechtslund, & M. Sandoval (Eds.), *Internet and surveillance: The challenges of web 2.0 and social media* (pp. 71–88). New York: Routledge.

Bail, C. A. (2014). The cultural environment: Measuring culture with big data. *Theory and Society*, *43*(3), 465–482.

Barry, L., & Fisher, E. (2019). Digital audiences and the deconstruction of the collective. *Subjectivity*, *12*: 210–227.

Baudrillard, J. (1981). *For a critique of the political economy of the sign.* St. Louis: Telos Press.

Beer, D. (2009). Power through the algorithm? Participatory web cultures and the technological unconscious. *New Media and Society*, *11*(6), 985–1002.

Beer, D. (2019). *The data gaze: Capitalism, power and perception.* London: Sage.

Bell, D. (1999). *The coming of post-industrial society: A venture in social forecasting.* New York: Basic Books.

Bilic, P. (2016). Search algorithms, hidden labour and information control. *Big Data & Society*, *3*(1), 341–366.

Birchall, C. (2016). Shareveillance: Subjectivity between open and closed data. *Big Data & Society*, *3*(2), 1–12.

Borgmann, A. (1999). *Holding on to reality: The nature of information at the turn of the millennium.* Chicago: University of Chicago Press.

Bucher, T. (2012). Want to be on the top? Algorithmic power and the threat of invisibility on Facebook. *New Media and Society*, *14*(7), 1164–1180.

Bucher, T. (2016). The algorithmic imaginary: Exploring the ordinary affects of Facebook algorithms. *Information, Communication & Society*, *4462*(April), 1–15.

Castells, M. (2010). *The information age: Economy, society and culture.* Oxford: Wiley-Blackwell.

Cheney-Lippold, J. (2011). A new algorithmic identity: Soft biopolitics and the modulation of control. *Theory, Culture and Society*, *28*(6), 164–181.

Cohn, J. (2019). *The burden of choice: Recommendations, subversion, and algorithmic culture.* New Bruswick: Rutgers University Press.

Couldry, N., & Turow, J. (2014). Advertising, big data, and the clearance of the public realm: Marketers' new approaches to the content subsidy. *International Journal of Communication*, 8, 1710–1726.

Crawford, K. (2016). Can an algorithm be agonistic? Ten scenes about living in calculated publics. *Science, Technology & Human Values*, *41*(1), 77–92.

Danaher, J., Hogan, M. J., Noone, C., Kennedy, R., Behan, A., De Paor, A., Felzmann, H., Haklay, M., Ming Khoo, S., Morison, J., Helen Murphy, M., O'Brolchain, N., Schafer, B., & Shankar, K. (2017). Algorithmic governance: Developing a research agenda through the power of collective intelligence. *Big Data and Society*, *4*(2).

Derrida, J. (1974). *Of grammatology*. Baltimore: Johns Hopkins University Press.

Dijck Van, J. (2014). Datafication, dataism and datavcillance: Big data between scientific paradigm and ideology. *Surveillance and Society*, *12*(2), 197–208.

Dobson, J. E. (2019). *Critical digital humanities: The search for a methodology*. Chicago: University of Illinois Press.

Ellul, J. (1964). *The technological society*. New york: Knopf.

Feenberg, A. (1991). *Critical theory of technology*. New York: Oxford University Press.

Ferguson, A. (2017). *The rise of big data policing: Surveillance, race, and the future of law enforcement*. New York, NY: New York University Press.

Fisher, E. (2010). *Media and new capitalism in the digital age: The spirit of networks*. New York: Palgrave.

Fisher, E. (2020). Can algorithmic knowledge about the self be critical? Stoccheti, M. (Ed), The digital age and its discontents, Helsinki: Helsinki University Press, 111–122.

Fisher, E., & Fuchs, C. (Eds.). (2015). *Reconsidering value and labour in the digital age*. New York: Palgrave.

Fisher, E., & Mehozay, Y. (2019). How algorithms see their audience: Media epistemes and the changing conception of the individual. *Media, Culture and Society*, 41(8), 1176–1191.

Foucault, M. (1977). *Discipline and punish: The birth of the prison*. London: Vintage Books.

Foucault, M. (1994). *The order of things: An archeology of human sciences*. New York: Vintage Books.

Foucault, M. (2006). *The hermeneutics of the subject: Lectures at the collège de france, 1981–1982*. New York: Palgrave Macmillan.

Fuchs, C. (2011a). An alternative view of privacy on Facebook. *Information*, *2*(1), 140–165.

Fuchs, C. (2012). The political economy of privacy on Facebook, *Television and New Media*, 13(2), 139–159.

Gillespie, T. (2012). Can an algorithm be wrong? *Limn*, 2, 21–24.

Gillespie, T. (2014). The relevance of algorithms. In Tarleton Gillespie, Pablo J. Boczkowski, and Kirsten A. Foot (Eds.), *Media Technologies: Essays on Communication, Materiality, and Society*, pp. 167–194.

Gillespie, T. (2016). Trendingistrending: When algorithms become culture. In Robert Seyfert and Jonathan Roberge (Eds.), *Algorithmic Cultures: Essays on Meaning, Performance and New Technologies*, pp. 52–75.

Gitelman, L. (Ed.). (2013). *Raw data is an oxymoron.* Cambridge, MA: MIT Press.

Grosser, B. (2017). Tracing you: How transparent surveillance reveals a desire for visibility. *Big Data and Society* 4(1), 1–6.

Habermas, J. (1972). *Knowledge and human interests.* Boston: Beacon Press.

Habermas, J. (1985). *The theory of communicative action.* Boston: Beacon Press.

Habermas, J. (1991). *The structural transformation of the public sphere: An inquiry into a category of bourgeois society.* Cambridge, MA: Mit Press.

Hallinan, B., & Striphas, T. (2014). Recommended for you: The Netflix prize and the production of algorithmic culture. *New Media and Society, 18*(1), 117–137.

Heidegger, M. (1977a). The question concerning technology. In M. Heidegger (Ed.), *The question concerning technology and other essays* (pp. 3–35). New York: Hatper Torchbooks.

Heidegger, M. (1977b). *The question concerning technology and other essays.* New York: Hatper Torchbooks.

Hildebrandt, M. (2019). Privacy as protection of the incomputable self: From agnostic to agonistic machine learning. *Theoretical Inquiries of Law, 20*(1), 83–121.

Jameson, F. (1981). *The political unconscious: Narrative as a socially symbolic act.* Ithaca, NY: Cornell University Press.

Kant, I. (1999). *Critique of pure reason.* Cambridge: Cambridge University Press.

Kant, I. (2015). *Critique of practical reason.* Cambridge: Cambridge University Press.

Kennedy, D. (2011). Industrial society: Requiem for a concept. *The American Sociologist, 42*(4), 368–383.

Kennedy, H., & Moss, G. (2015). Known or knowing publics? Social media data mining and the question of public agency. *Big Data & Society, 2*(2), 205395171561114.

Kim, P. (2017). Auditing algorithms for discrimination. *University of Pennsylvania Law Review Online, 166*(1), 10.

Kitchin, R. (2017). Thinking critically about and researching algorithms. *Information, Communication & Society, 20*(1), 14–29.

Knorr Cetina, K. (1999). *Epistemic cultures: How the sciences make knowledge.* Cambridge, MA: Harvard University Press.

Levenberg, L., Neilson, T., & Rheams, D. (Eds.). (2018). *Research methods for the digital humanities.* New York: Palgrave.

Lyotard, J.-F. (1984). *The postmodern condition: a report on knowledge.* Minneapolis, MN: University of Minnesota Press.

Mackenzie, A. (2005). The performativity of code software and cultures of circulation. *Theory, Culture & Society, 22*(1), 71–92.

Mackenzie, A. (2013). Programming subjects in the regime of anticipation: Software studies and subjectivity. *Subjectivity, 6*(4), 391–405.

Mackenzie, A., & Vurdubakis, T. (2011). Code and codings in crisis: Signification, performativity and excess. *Theory, Culture & Society, 28*(6), 3–23.

Mager, A. (2012). Algorithmic ideology: How capitalist society shapes search engines. *Communication & Society, 15*(5), 769–787.

Mager, A. (2014). Defining algorithmic ideology: Using ideology critique to scrutinize corporate search engines. *TripleC: Communication, Capitalism and Critique, 12*(1), 28–39.

Marres, N. (2017). *Digital sociology: The reinvention of social research.* Cambridge: Polity.

Marx, K. (1992). *Capital: Volume 1: A critique of political economy.* New York: Peguin.

Mattelart, A. (2003). *The information society: An introduction.* Thousand Oaks, CA: Sage.

Mayer-Schönber, V., & Cukier, K. (2013). *Big data: A revolution that will transform how we live, work, and think.* New York: Houghton Mifflin Harcourt.

McCarthy, T. (1988). Introduction. In *On the logic of the social sciences* (pp. vii–x). Cambridge, MA: MIT Press.

McQuillan, D. (2015). Algorithmic states of exception. *European Journal of Cultural Studies, 18*(4–5), 564–576.

Mehozay, Y., & Fisher, E. (2018). The epistemology of algorithmic risk assessment and the path towards a non-penology penology. *Punishment and Society, 21*(5), 523–541.

Mittelstadt, B. (2016). Auditing for transparency in content personalization systems. *International Journal of Communication, 10*, 4991–5002.

Neyland, D. (2015). On organizing algorithms. *Theory, Culture and Society, 32*(1), 119–132.

Ortner, S. B. (2005). Subjectivity and cultural critique. *Anthropological Theory, 5*(1), 31–52.

Pariser, E. (2012). *The filter bubble: How the new personalized web is changing what we read and how we think.* New York: Penguin Books.

Parsons, T. (1968). *The structure of social action: With a new introduction, Vol. 1.* New York: The Free Press.

Pasquale, F. (2015a). The algorithmic self. *The Hedgehog Review, 17*(1), 1–7.

Pasquale, F. (2015b). *The black box society: The secret algorithmic that control money and information.* Cambridge: Harvard University Press.

Postman, N. (1993). *Technopoly: The surrender of culture to technology.* New York: Vintage Books.

Rebughini, P. (2014). Subject, subjectivity, subjectivation. *Sociopedia, ISA.*

Robins, K., & Webster, F. (1999). *Times of technoculture: From the information society to the virtual life.* London: Routledge.

Rona-Tas, A. (2020). Predicting the future: Art and algorithms. *Socio-Economic Review, 18*(3), 893–911.

Rouvroy, A., & Stiegler, B. (2016). The digital regime of truth: From the algorithmic governmentality to a new rule of law. *Online Journal of Philosophy, 3*, 6–29.

Ruckenstein, M., & Schüll, N. D. (2017). The datafication of health. *Annual Review of Anthropology*, 46, 261–278.

Sauter, T. (2013). Whats on your mind? Writing on Facebook as a tool for self-formation. *New Media and Society, 16*(5), 1–17.

Soll, J. (2014). *The reckoning: Financial accountability and the rise and fall of nations.* New York: Basic Books.

Stehr, N. (2001). *The fragility of modern societies: Knowledge and risk in the information age.* Thousand Oaks, CA: Sage.

Striphas, T. (2015). Algorithmic culture. *European Journal of Cultural Studies, 18*(4–5), 395–412.

Thrift, N. (2005). *Knowing capitalism.* London: Sage.

Touraine, A. (1971). *The post-industrial society: Tomorrow's social history: classes, conflict and culture in the programmed society.* New York: Random House.

Tufekci, Z. (2019). How recommendation algorithms run the world. *Wired,* April 22.

Turow, J. (2011). *The daily you: How the new advertising industry is defining your identity and your worth.* New Haven: Yale University Press.

Turow, J., & Draper, N. (2014). Industry conceptions of audience in the digital space. *Cultural Studies, 28*(4), 643–656.

Van Dijck, J., & Poell, T. (2016). Understanding the promises and premises of online health platforms. *Big Data & Society.*(June).

Veltri, G. A. (2019). *Digital social research.* Cambridge: Polity.

Webster, F. (2002). *Theories of the information society.* London: Routledge.

Winner, L. (1977). *Autonomous technology: Technics-out-of-control as a theme in political thought.* Cambridge, MA: MIT Press.

2 How algorithms think about humans?

With Yoav Mehozay

Media epistemes and the changing conception of the individual

Recent years have witnessed the emergence of an expanding array of socio-technical apparatuses promising to transform big data into knowledge with the aid of algorithms. This huge effort to translate data into knowledge – involving varied bodies of knowledge, practices, methodologies, technologies, practitioners, and discourses, and requiring huge capital investments – is taking place in diverse fields. Digital media are arguably at the forefront of this effort. From social networking sites to online vendors, digital media are a particularly suitable breeding ground for the rise of algorithmic knowledge, for at least three reasons. First, being interactive, digital platforms allow for the incessant monitoring and accumulation of big data, particularly data representing users' behavior (Zuboff, 2015). Second, the big promise of digital media is personalization, or mass-customization of content. Digital media sites are therefore heavily invested in creating knowledge about users from their data in order to effectively deliver on the promise of personalization (van Dijck et al., 2018). And third, as digital media are (mostly) free to use, their business model is based predominantly on their ability to commodify data by turning them into knowledge and selling them to content producers and advertisers (Fisher, 2018).

The result has been a huge and complex apparatus – which we call here "algorithmic" as shorthand but which comprises a plethora of technologies, platforms, bodies of knowledge and procedures – designed to translate user-generated data into knowledge about users (van Dijck, 2013). This apparatus is part and parcel of almost any type of digital media. In some cases, it operates to *improve functionality*, for instance, by helping refine content recommendations for users (e.g.,

DOI: 10.4324/9781003196563-3

recommendations for further reading for readers of newspaper websites, or product suggestions offered by retailers). In other cases, algorithmic knowledge is the *core functionality* of the service. This is the case, for example, in dating sites that match partners based on personal data (Rudder, 2015). Whether in order to deliver personalized information or selling quality users' information to third parties, there is a great push to invest in algorithmic apparatuses, which strive to know platforms' users.

Algorithms can be said to "see" media users through data. This is apparent in all sorts of recommendation, sorting, and curation engines. Facebook, for example, needs to produce millions of unique, personalized, real-time newsfeeds every second. Users could not possibly see every post by every one of their Facebook friends (not to speak of advertisements). Indeed, every user sees only some of their friends' posts, based on algorithmic decision making. The motivations for this procedure can be multiple and even contradictory: keeping users on the website for as long as possible, creating as much engagement as possible, promoting one particular type of message over another, and so forth. But for the purpose of our argument here, these varied motivations make no difference, as they all benefit from gaining an intimate and accurate knowledge about users as possible. Regardless of why it does so, then, Facebook is keen to translate data into knowledge about every single member of its audience: their interests, tastes, biases, wishes, and so forth.

Of course, there is nothing new about the media seeking to know their audience. The development of the mass media in the 20th century was accompanied by the rise of techniques designed to uncover the whims and dispositions of readers, listeners, and viewers of the mass media outlets of the time – techniques that are now being revolutionized by the rise of big data and algorithms. Revenue from advertising was, then as now, the ultimate rationale of any commercial medium. Yet we cannot understand the emergence of the algorithmic gaze on the audience only by analyzing the changing field of advertising in the digital media era (Turow, 2011; Turow & Draper, 2014). The new algorithmic apparatus signals not simply a new means with which to pursue old ends more efficiently, but a completely new way of knowing, which tells us something about how human beings are perceived and conceptualized in an era saturated with algorithmic devices (Cheney-Lippold, 2011). In other words, underlying digital media is a new epistemology, a new way to know human-beings.

By recalling the history of audience measurement through the 20th century and locating the latest algorithmic episteme as part of that history, we ponder the nature of different epistemes, specifically, *their underlying assumptions concerning the nature of human beings.* What interests us is how new techniques by which the media measure their audience also point to a new way of thinking about humans. The knowledge that such technological systems create is performative (Callon, 2007): they not only assume what humans are, but act upon these assumptions, helping create that which they presume to measure. The question we pose is thus not whether the algorithmic episteme of digital media is more accurate and refined than the epistemic model used by the mass media. Instead, we want to unravel how this new episteme differs from its predecessor, and the ramifications of this difference for our understanding of the self.

The chapter proceeds by recalling how the mass media measured its audience during most of the 20th century. For simplicity's sake, we outline the overarching principles, which dominated the mass media's gaze at the audience, plastering over differences between different media forms, eras, and locations. We then follow the same analytical procedure concerning digital media and the algorithmic episteme underlying how they see their audience. Our purpose is to highlight the assumptions regarding human beings, which inform the algorithmic episteme.

Two methodological clarifications concerning periodization and conceptualization are in order. We distinguish two media systems, which dominate different historical periods, and highlight their respective, mutually exclusive techniques for measuring the audience. This ideal-typical presentation is meant for analytical clarity. Empirically, the picture is obviously much more complex and brings together all sorts of juxtapositions: between different kinds of media and different kinds of techniques. A movie produced and distributed by Netflix, for example, is a complex object involving not merely both mass and new media forms, but may also involve traditional and new techniques for measuring and identifying the audience. This means that the two epistemes do not exist in separated spheres of time and space, but interact and create new epistemic forms that need to be studied empirically (Kotliar, 2020). Hence, to speak of these two epistemes as distinct, and existing in different periods or distinct media spaces is justified only analytically and is meant to highlight the novelty of a new episteme.

The second methodological clarification concerns the notion of episteme. We adopt Foucault's terminology in order to draw a link between the empirical, local, and concrete, on the one hand, and the abstract, social, and structural, on the other hand. While we are dealing with concrete techniques and frameworks that are applied in a specific context, and geared toward producing knowledge for a limited purpose – that is, media – we want to suggest that underlying this particular case is a much broader, societal episteme, which manifests itself in multiple and diverse fields.

Seeing the audience in the mass media: The scientific episteme

The era of the mass media that dominated the 20th century – particularly film, radio, television, newspapers, and books – was the first, where, on a grand scale, content producers were not in the presence of their audience. Unlike musicians in a concert hall, storytellers in a room, or actors on a theatre stage, producers of mass media had no means of directly seeing their audience, determining who they were, and observing how people responded to the content conveyed. This disjuncture between the media and their audience gave rise to an epistemic project that grew parallel to the development of the mass media: that of identifying and characterizing the audience for each broadcast form, platform, and content.

For a brief period, these questions were answered intuitively by media practitioners who would hold "imaginary interlocutors" or "reference persons" in their heads while producing media content such as journalistic accounts (Pool & Schulman, 1959). Otherwise, the mass media simply assumed a coherent, homogeneous crowd of people. The audience was conceived as comprised of a multiplication of a similar, standard, universal person (in practice, an upper-middle-class, college-educated, white, urban male). This intuitive, prescientific episteme assumed a singular, rational individual. Under such an assumption, audience research would have seemed redundant, and was indeed virtually nonexistent. Assuming a universal receiver, research was limited mainly to defining and measuring the effectiveness of messages.

But this prehistory soon gave way to a new understanding that would dominate the project of deciphering the mass media audience in the 20th century: a conception of the audience as a heterogeneous multitude, comprising diverse groups with different characters, tastes, and desires. These groups needed to be identified more intimately in order to construct and address messages more accurately. To achieve that, the audience needed to be studied more carefully and systematically (Buzzard, 2012, pp. 3, 13ff). The dominant episteme for seeking the audience that grew out of this effort stands on two foundations: social theory and empirical research. This project drew heavily on the epistemology of the social sciences (Ettema & Whitney, 1994, p. 9), and as such can be characterized as scientific. The key event in this process was the rise of communication studies in the first half of the 20th century, with its focus on studying public opinion, and its close ties with the media industry (Katz, 2009; Pooley & Katz, 2008). Pioneers of communication studies came from academia and imported a scientific framework to tackle the question of how to see the audience. The fruits of this scientific endeavor can be called the *scientific episteme.*

By denoting this episteme "scientific", we do not mean to decree whether audience research done by the media industries is worthy of the term or not. There are two important qualifications here. First, such research drew on a very specific strand of theory and methods associated with administrative research, rather than critical one (Horkheimer, 2002; see also Lazarsfeld, 1941). And second, this kind of research can be said to be scientific in form but not in substance and spirit. Most of its output was not governed by an academic ethos, such as producing knowledge for knowledge's sake, or making findings public and part of a scientific discourse. Rather, it was aimed to advance extra-scientific objectives, such as increasing the efficiency of messages, or enlarging the circle of audience.

In building the scientific episteme, sociological and cultural theories served as templates to outline the contours and central coordinates for analyzing the audience. This led to an understanding of the audience as comprising various groups based on *social categories*, such as sex, race, income, and education. Drawing on these theoretical templates, academics and practitioners engaged in intensive empirical research, probing audiences with the aid of surveys, indepth interviews, and mathematical tools for measuring and analyzing ratings (Buzzard, 2012; Napoli, 2010). The scientific episteme even led to innovations in research methods, making significant contributions to the development of focus groups as a tool to study audience perceptions of media messages (Lunt & Livingstone, 1996). The combination of social theory and empirical research meant that the mass media could learn about their audience based on very small samples. That is, the scientific episteme allowed seeing each case, or each individual, as a representative of a larger whole – the particular social category of which she is part. Samples of a few hundred or even a few dozen cases could provide information about a whole social system – a particular audience population in this case.

Underlying this episteme was an *ascriptive* conception of the individual: each individual could be assigned to a category, which could then be sociologically and culturally characterized. For example, the-ories that infer a high correlation between social categories and cul-tural taste, from Weber's (1958) to Bourdieu's (1984), informed practices of audience segmentation. Using the scientific episteme, the mass media thus assumed – and constructed – an audience comprising different, but internally homogenous, social categories. It then tailored content for each category, based either on members' presumed inter-ests or on the demands of advertisers.

The scientific episteme created a huge body of knowledge con-cerning different audiences. The mass media were attentive to this new knowledge, and the media landscape changed. For example, whereas the mass media once produced magazines of general interest aimed at citizens, now dozens of categories of magazines were produced to be read by identifiable groups, or segments, of the broader audience, which became diversified with time: middle-class, white, suburban housewives; upper-middle-class minority professional women; and so forth. Thus, magazines became increasingly gendered, age-based, and appealing to various lifestyles. In synchronic media, such as network television, segmentation could be achieved through time management. For example, soap operas (targeting housewives) tended to be broadcasted in the early morning, while financial reporting (targeting male professionals) tended to be broadcasted in the evening.

The mass media not only discovered different audiences but also created them, in the way that any discourse creates, or interpellates, the objects it identifies (Althusser, 1970; Foucault, 1994). Media content was produced to appeal to specific social categories – and not others – based on the scientific episteme. Thus, for example, where gender was perceived to be a binary category, producers responded by creating binary media content. As content designated for certain groups became available, individuals in those groups were more likely to consume that designated content, and take on the identity assigned to them. For example, the media industry identified young, urban, professional white males as a specific social category – an audience segment – and began offering men's magazines. These magazines appealed to enough individuals in that category so that content continued to be tailored to the perceived interests and concerns of this group (Stevenson et al., 2001). It is in this sense that we can speak of the media not merely as measuring and seeing the audience but also as projecting this image on the audience and manufacturing it (Bermejo, 2009).

To be sure, this way of seeing the mass media audience – as a set of segments or categories – was motivated by marketers' need to find (or construct) targeted audiences for specific advertising messages. It is in this sense that Ien Ang argues that the audience is "an imaginary entity ... constructed from the vantage point of the institutions, in the interests of the institutions" (Ang, 1991, p. 2). Likewise, Philip Napoli speaks of "institutionally effective audiences" as "those that can be efficiently integrated into the economics of media industries" (Napoli, 2010, p. 3). However, as noted earlier, understanding media conceptions of the audience merely from the point of view of the political economy of commercial media and its need to sell advertisements falls short of explaining why audience segmentation took the form it did. This is because, regardless of the motivations behind them, even these institutionally motivated constructions were not made up out of thin air but were based on real social categories: theoretically informed, empirically tested, and relying on small representative samples.

To sum up, the conception of the audience that developed in the mass media was based on a scientific episteme, whereby individuals were catalogued into social categories. Such audience making, as Ettema and Whitney (1994) call it, is not "the assemblage of individual readers, viewers, or listeners who receive messages". In the mass media era, members of the audience have no individual existence; rather, they are conceptualized as part of "institutionally effective audiences that have social meaning and/or economic value within the system" (p. 5). Institutional audiences are "generated by research services", and "include specialized or segmented audiences whose particular interests are anticipated – or created – and then met by content producers" (p. 5). The scientific episteme, the mode by which knowing the audience was made possible in the mass media era, is therefore central to these constructions. It informs the tools by which the audience is seen. As Ang puts it eloquently, "through audience measurement, the [media] has equipped itself with a basic mechanism to get to know the audience in a way that suits the industry's interests" (Ang, 1991, p. 3).

Seeing the audience in digital media: The algorithmic episteme

The rise of digital media was accompanied by the development of a new episteme, a new way of knowing the audience, one centered on big data and algorithms (Antique, 2017). This transformation was not revolutionary by any means: it did not take place overnight and it did not topple previous epistemes (Kotliar, 2020). The result was a layered transformation, one that can be illustrated with reference to the advertising industry, a key player in the media's drive to know its audience. As advertising first moved into the Web in the mid-1990s it imported the set of practices developed during almost a century of mass media advertising. But as the advertising industry began to realize the unique qualities of the digital ecosystem, it began toying with new techniques, which eventually turned into a whole new advertising paradigm (Fisher, 2015, 2017).

In the present digital media environment, we can therefore identify a novel paradigm for seeing the audience, one that is not an extension, improvement or development of previous modes but something completely new (Buzzard, 2012, p. 10). Three socio-technological features distinguish digital media from the mass media and underlie the new paradigm for seeing the audience, which we call the algorithmic episteme:

I User-generated data

One distinct feature of digital platform compared with the mass media is their newfound ability to monitor individuals' data, created as user encounter digital objects, both online and offline. We use the term data in its broadest sense to cover any digital record of behavior: a post on Facebook (i.e., information), the sum of money spent on a credit card purchase (data in the strict sense), or the amount of time spent reading a news story online (meta-data). Technological components which allow user interactivity, such as JavaScript, AJAX, or wiki, as well as digital media installed in credit cards and vehicle monitoring devices, or the signals sent from cell phones to transmission towers – all help in collecting and storing the digital footprints that users produce as they live their lives, almost invariably interwoven with digital media. Some data, particularly content, is produced by users in the immense spaces of social media and Web 2.0 platforms, which are specifically designed for social communication among users.

II Inter-connected platforms

This carnival of data production takes place within a structured and interconnected media ecosystem. Speech-acts and data points can be traced back to users, and users' data is monitored not merely per a particular website, but by an increasingly ominous techno-social constellation allowing for a "culture of connectivity" (van Dijck, 2013). Cloud computing, cross-platform surveillance, and monopolization have given rise to a historically unprecedented database in size and in type – that is, big data. Big data serves as a raw material used to enhance the performance of a given platform and create information commodities, sold mostly to advertisers (Zuboff, 2020).

III Algorithms

Algorithms are the key technology used in rendering data into valuable information and knowledge. In recent years, the field of algorithms has been heavily influenced by big data, turning algorithms into pattern discovery technologies, capable of learning and self-adjusting dynamically in response to a flow of new data. Building on mathematical and technical advances such as machine learning, deep learning, artificial intelligence, and neural networks, algorithms can be now used as decision-making devices (Mackenzie, 2017).

While the goals for seeking the audience in digital media compared with the mass media remained much the same, the new media environment called for a whole new episteme to achieve them. In contrast to the mass media, digital media are characterized less by production of content and more by mediation between content producers and content consumers. Digital platforms serve predominantly as personalized curators and mediators, which trawl through a large supply of content and match it to users' personal wants, interests, and likes. Digital media, then, are largely driven by audience members addressing each other. Hence, to ask "how the audience is seen by media producers" gained another meaning, referring to how media produsers (Bruns, 2008) (e.g., users writing Tweets) imagine their audience, that is, other users (Litt & Hargittai, 2016; Marwick & Boyd, 2010). We, however, refer here only to how digital platforms imagine their users, as this involves the novelty of an algorithmic episteme.

Compared with the mass media, the need to see the audience becomes ever more crucial with digital media, since the latter are fundamentally about *personalization* (Buzzard, 2012, p. 5). The promise inherent in almost all digital media is precisely the promise to deliver personalized information quickly and accurately, marrying content to users (Athique 2017). Moreover, it is a promise to personalize content seamlessly from the point of view of users, based on a careful study of individual wants and interests. This has been referred to as a shift from search to discovery (Sadeh, 2015): *search* implying a human subject who knows what she is looking for, *discovery* implying serendipitously running into content offered by the media platform based on how it sees her and interprets her wants. For digital media platforms, such as content discovery platforms Outbrain and Taboola, or dating website OkCupid, knowing their audience individually is part and parcel of the service, in a way that cannot really be compared to traditional content delivery by the mass media.

The other motivation for media platforms to know their audience – targeted advertising – persists. The underlying political economy, too, remains virtually the same: delivering free information to the audience and targeting that audience as consumers (Lee, 2011; McGuigan & Manzerolle, 2014). Yet again, while this motivation can be said to ultimately hark back to financial gains, it fails to explain *how* knowledge about the audience is collected and analyzed. And so we are left with the need to explain not the constant ends, but the changing means. What has changed is the (assumed) ability of the media to know their audience more intimately than ever before, by turning users' engagement, interactivity, or mere use of digital media into data.

Turning data into self: How algorithms see people

That lots of personalized data is produced by users, accumulated by platforms, and made available to algorithms tells us little about how algorithms see us. How are these huge quantities of data turned into knowledge about human qualities or wants? Assuming that Spotify has a huge set of data about its users' listening history, plus any conceivable data available from third-party resources, how are all these data, or input in algorithmic terms, rendered into musical recommendations, or output? To open a window onto the assumptions underlying the algorithmic episteme, let us consider the involvement of digital media in the construction of the quantified self. The quantified self concerns a conscious and voluntary individual attempt to mobilize the algorithmic episteme in order to create a new type of knowledge about the self and for the self (Lupton, 2016; Neff & Nafus, 2016). Let's take a close look at People Keeper as a simplified case of algorithmic knowledge. People Keeper is a technological system, comprising a smart watch and an online application, aimed at helping users "optimize their social lives" (Keeper, 2018). It collects numerous data points, both automatically and through self-reporting, concerning interactions with other people, and yields a coherent outcome about how users feel about another person. A promotional video for People Keeper explains:

We only have so much emotional bandwidth and limited time. Our social circles are widening. All these relationships can be overwhelming. Now there's an app. People Keeper tracks your physical and emotional response while you're hanging out, then analyzes the data to identify who stresses you out, makes you excited, sad, or happy. See how your relationships stack up, and let People Keeper find the ones that work for you. It will automatically manage your relationship, so you don't have to, scheduling time with people that make you feel good, and blocking the ones that don't. Forget fake friends, failed romance, and FOMO. Optimize your social life with People Keeper. (Keeper, 2018)

This is an extreme example, as People Keeper is not primarily a consumer product but an art project created by artists Lauren McCarthy and Kyle McDonald "to explore the implications of quantified living for relationships" (Groden, 2015). Though a tongue-in-cheek, People Keeper can nevertheless tell us a lot about the promise of the algorithmic episteme, or the contours of its utopian – as well as dystopian – horizons. The imagining of a technological system is at the same time also a work-plan for engineers, designers, and entrepreneurs, setting the direction for further development (Flichy, 2007). It is reacting and commenting on a technological reality as well as participating in its constitution.

Using People Keeper as a vignette for the algorithmic imaginary (Bucher, 2016), let's uncover a few tenets of the algorithmic episteme. The most fundamental assumption of the algorithmic episteme is that to know the audience is to know their digitally registered behavior; that is, that individuals can be understood through data. The individual is known through the pattern that emerges from the data she produces, data indicating behavior – either conscious and oriented or unconscious and assumed. "Behavior" in this case is something of a misnomer, as it refers to anything that can be digitally monitored, from the user's navigation on the web through movements of her body in space to even deeper layers, such as the firing of neurons in her brain. Behavior, in other words, is anything that has some *objective* digital footprint. This data universe excludes subjective and internal processes that cannot be datafied. The most immediate behavioral data in digital media concerns behavior within the platform itself, involving mostly content consumption. Digital media, then, see their audiences primarily in line with their behavior on the platform.

The algorithmic episteme gives primacy – even exclusivity – to surface over depth in its conception of individuals. To *know* someone does not mean to analytically and empirically understand the reasons for her behavior, but instead to be able to recognize patterns of behavior. This excludes any attempt at a sociological, psychological, or indeed any theoretical etiology as a tool for understanding. It also signals a rejection of essentialism in how we think about people; if there is no underlying foundation for who we are but only surface behavior, it becomes impossible to speak of any individual as representing or existing within a larger systemic whole, such as gender or class. In other words, the algorithmic episteme assumes no social ascription to the audience of digital media.

Excluding the etiology of behavior also expunges the notion of interpretive understanding (*Verstehen*, in Weber's terms [(Weber, 1921)]) from how algorithms see individuals. To some extent, subjective interpretation and hermeneutic reasoning are seen as a hindrance to self-understanding. Reflexive and narrativistic models of knowing need to be overcome in order to get to the building blocks of the self: a heavy load of data points representing discrete digital events.

Seeking the mass media audience involved knowing individuals in terms of their location in a social matrix. This matrix could be quite complicated and include multiple variables. For example, an individual could be located under the rubric of male/Latino/ urban/ gay/ middle-class. Assuming five variables, as in this example, and three values for each variable would result in 5^3 (125) rubrics under which individuals could be catalogued. The algorithmic episteme puts us in a completely different numerical universe, with possibly hundreds of variables and hundreds of values for each variable. To the extent that such data could be rendered in natural language (e.g., via a table), it would contain thousands upon thousands of rubrics, making it humanly impossible to process.

But the rubrics in the algorithmic episteme are not only quantitatively larger than the categories used by the scientific episteme of the mass media, they are also qualitatively different: they represent not social categories, but patterns of data. Since the algorithmic episteme allows for no theory and no assumptions regarding deeper causes of human behavior, no *a priori* variables can be deduced for analysis. Hence, the algorithmic episteme has an omnivorous approach to variables: any variable can be added to the mix as there is virtually hardly a technical limit to the number of variables that can be processed (Mayer-Schönber & Cukier, 2013). As algorithmic knowledge is increasingly aided by machine learning, neural networks, and artificial intelligence, the need to control variables and even the need for supervised learning is reduced. With no theory of the self, no essentialist assumptions, and no attention to etiological causes, digital media platforms amass all the variables they can, making no judgments concerning the relevance of these variables to a question at hand.

Algorithmic knowledge about the audience is based on what Deborah Lupton calls "lively data" (Lupton, 2016, p. 42ff). The "lively" quality of data refers to both the vitality and dynamism of data – the fact that they are incessantly created and flow – and to the fact that data emerges out of – and constitutes an unmediated representation of – "life itself", that is, aspects of life which are now digitally registered – affective, communicative, relational, emotional, erotic, and so forth, as well as the most mundane, trivial aspects of "humanness", such as the brand of our mobile device, or the duration it takes us to read a news story. This constant data flow underscores the assumption of humans as dynamic and open-ended entities rather than essentialist: there is always new data, and so the image of a person in the eyes of the algorithm is always open for change.

In this epistemic universe, the guiding measure for successfully seeing the audience is predictive, rather than explanatory (Mackenzie, 2015). To know the audience is to be able to predict their future behavior. Measuring success is done by posing *posteriori* tests, which have practical, rather than theoretical grounding. This is known as A/B testing, and it involves human decision making, which lies outside the internal logic of algorithms. For example, for Netflix's or Amazon's recommendation engines, a predictive algorithm (B) can be said to be successful to the extent that it delivers more clicks than another algorithm (Ai) or a random recommendation (Aii). The algorithmic episteme makes no claims for truth, but rather for function. It cannot even be said to be wrong in any theoretical or mathematical sense. As Lowrie puts it, algorithms

> can only be evaluated in their functioning as components of extended computational assemblages; on their own, they are inert. As a consequence, the epistemological coding proper to this evaluation does not turn on truth and falsehood but rather on the efficiency of a given algorithmic assemblage. (Lowrie, 2017, p. 1)

Rather than thinking about individuals as part of a social category, or characterizing them by their qualities, the algorithmic episteme instead thinks about them in terms of patterns of behavior, or data patterns.

Since individuals could fit into thousands of rubrics, the algorithmic episteme can do away with a conception of individuals as belonging to a binary (man/woman) or otherwise discrete (lower-, middle-, upper-class) demographic category, contributing to what Richard Rogers has called a "post-demographic" conception of individuals (Rogers, 2009, Ch. 7). Such conception is indifferent to ascriptive social categories (such as gender or income), and indifferent to the master-narratives of modernity (such as nationalism and class). Instead it offers a much more nuanced gaze of a self, defined and characterized by the pattern of data it produces. Such conception of the self presumably requires no theory. The immense quantity and variety of big data helps make the leap from actual empirical phenomena (as if data were indeed a paragon of reality) to knowledge without the need for abstracting, theorizing, modeling, and hypothesizing.

What puts individuals together in the same category – a category for which an actuarial analysis can be performed, and the future behavior of which can be predicted – is merely a similar pattern of data points. Here, then, the social and political ramifications of the divergence between the scientific episteme and algorithmic episteme for knowing the audience becomes apparent. The mass media viewed the audience in terms which could be put in *natural language*: two people could be conceived as similar (e.g., likely to share a similar fondness for TV drama) to the extent that they were both educated, upper-class, and male. In contrast, nothing can be said about the similarities of two individuals sharing the same behavioral pattern in digital media, except that they fall into a similar predictive category. Consider two individual readers of *The New York Times* receiving similar recommendations for further reading from Outbrain, or similar recommendations for books on Amazon. The only thing we can know for sure about these two individuals is that they share a data pattern. But since we do not even know what variables were included in the algorithms that produced this data pattern, let alone the coefficient of each, we cannot translate this similarity in data patterns into natural language pertaining to social categories.

Conclusion: Toward a post-social conception of the individual

The increasing centrality of data and algorithms in the workings of digital media, specifically in how the media attempt to decipher who their users are and what they want, cannot be explained away solely in technical terms, as a result of newfound technological means.

Rather, as this chapter has shown, it entails a way of knowing the audience, which is radically different from how the mass media saw their audience in the past. Media outlets have always engaged in practices aimed at gauging their audiences in order to ensure their messages find receptive ears and eyes. Their motivations for this were twofold and mostly indistinguishable: communicative and commercial. With the rise of digital media, the means of knowing the audience – what we have called as shorthand an episteme – have changed radically. And as we have argued, this change in the episteme was accompanied by a parallel change in the conception of the individuals who make up that audience. As the chapter has shown, two epistemes for knowing the audience can be discerned: scientific and algorithmic. These are ideal types in the Weberian sense, but they can be associated with distinct media environments: the mass media and digital media, respectively.

Our central purpose has been to unravel the assumptions concerning *what humans are*, whish underlie and inform these distinct epistemes. We do not argue here for any causal link between media forms and the conception of individuals; rather, we aim to lay bare the assumptions behind these distinct epistemes, or the conditions that make them possible. What we have seen is a move from an ascriptive conception of individuals in the construction of the audience in the mass media to what might be called a performative conception of the audience in digital media. This entails seeing individuals based on the behavioral data they produce (Rouvroy & Stiegler, 2016), bypassing their self-understanding and identifying patterns from which a predictive behavioral analysis can be deduced (Barry & Fisher, 2019). We dub it "performative" to allude to the primacy, even exclusivity, it gives to how users perform online, what they do – either consciously and intentionally or not. Such a conception seeks surface behavior rather than unraveling deep structures; it foregoes any attempt to ground itself in, or offer any theory of, the self, and an etiology of behavior; lastly, it forgoes any concept of humans as sociological beings, as being a part of a social structure and belonging to social categories, and sees them instead as amalgams of ever-changing, dynamic, lively data-points.

Compared with the scientific episteme, the algorithmic episteme offers a very different idea concerning what *knowing* is. We can think about that difference in temporal terms: whereas the former is directed backwards in time, the latter is future-oriented. Rather than attempting to explain phenomena (e.g., a behavior) as a result of past or existing causes, the algorithmic episteme seeks to predict behavior.

This shift from a regime of truth to a regime of anticipation is growing increasingly dominant with predictive algorithms (Mackenzie, 2015). Knowledge, in the case of the algorithmic episteme, boils down to the ability to anticipate future trends and patterns.

The significance of theorizing this emerging episteme goes further than merely laying bare an emergent conception of the individual. These epistemes are performative: they not only assume a certain human being but also construct an individual, which they presume merely to measure or identify. As Donald MacKenzie (2007) elegantly puts it, these techno-social amalgams of technology, practitioners, bodies of knowledge, discourses, and so forth are "not a camera", which provides a snapshot of what human beings are, but "an engine" that partake in shaping them.

While algorithms work behind the backs of users and remain opaque to them (Pasquale, 2015), their ramifications are increasingly apparent. Users are well aware of the algorithmic nature of their interactions with digital technology; they know they are being watched and monitored. Users use their "algorithmic imagination" (Bucher, 2016) to see the content they are offered as indication of how they themselves are being seen. To some extent (albeit with critical distance), they also see it as an algorithmic reflection of their self. To change the metaphor offered earlier, the media, then, act here not as a camera, but as a mirror, reflecting back the image they capture. This, according to Gillespie, creates a feedback loop by which "the algorithmic presentation of publics back to themselves shapes a public's sense of itself" (Gillespie 2014).

Thus, the difference between the two ways of seeing the audience – in the mass media and digital media – is not merely epistemological; rather, the distinct conception of the individual, which underlies each has ramifications in terms of how the audience sees itself. The two epistemes reflect and reify distinct ontologies of the individual. The algorithmic episteme – embedded ultimately in a series of algorithms processing personal data in order to operate, for example, a recommendation engine in a website – is a "configuration through which users and/or clients are modeled and then encouraged to take up various positions in relation to the algorithm at work" (Neyland, 2015, p. 122). We must be careful not to see this transformation only in technical terms, as the result of the development of new technological tools, but as constructing what amounts to a new way of seeing ourselves.

To think of ourselves through the gaze of the epistemic episteme – that is, as the sum of the data patterns that our actions leave on digital platforms – may have sociological and political ramifications. The ascriptive conception of the self encouraged by the mass media was embedded within a *social* cosmos were individuals could see themselves as a part of a relatively limited, internally homogenous demographic categories. During most of the 20th century, identity was predominantly based on ascription to a category of people who are similar. By subscribing to an identity of "worker", "woman", or "black", individuals did not assume a totalistic identity between themselves and every other individual within the category. Instead, it assumed that individuals in the same category are identical in what was politically significant, for example, that they suffer from a categorical discrimination, or having similar material interests.

The performative conception of the individual, which underlies the algorithmic episteme, suggests the specter of a post-social, post-demographic cosmology. It assumes that identifying ourselves through ascription to a social category is too reductionist and instead offers categorizing individuals by their data patterns. This puts us at a risk of entering a post-social cosmos where individuals have a harder time identifying each other as sharing a similar category, as these categories remain opaque to us.

References

Althusser, L. (1970). Ideology and the ideological state apparatuses. In *A critical and cultural theory reader*. Toronto: University of Toronto Press.

Andrejevic, M. (2007). Surveillance in the digital enclosure. *The Communication Review, 10*, 295–317.

Ang, I. (1991). *Desperately seeking the audience*. London and New York: Routledge.

Athique, A. (2017). The dynamics and potentials of big data for audience research, *Media, Culture and Society, 40*(1), 59–74.

Barry, L., & Fisher, E. (2019). Digital audiences and the deconstruction of the collective. *Subjectivity 12*(3): 210–227.

Bermejo, F. (2009). Audience manufacture in historical perspective: From broadcasting to Google. *New Media and Society, 11*(1–2), 133–154.

Bourdieu, P. (1984). *Distinction: A social critique of the judgment of taste*. Cambridge: Harvard University Press.

Bruns, A. (2008). *Blogs, Wikipedia, second life, and beyond: From production to produsage*. New York: Peter Lang Publishing.

Bucher, T. (2016). The algorithmic imaginary: Exploring the ordinary affects of Facebook algorithms. *Information, Communication & Society, 4462*(April), 1–15.

Buzzard, K. (2012). *Tracking the audience: The ratings industry from analog to digital*. London: Routledge.

Callon, M. (2007). What does it mean to say that economic is performative? In Donald MacKenzie, Fabian Muniesa , and Lucia Siu (Eds.), *Do economists make markets?* Princeton: Princeton University Press, pp. 311–355.

Cheney-Lippold, J. (2011). A new algorithmic identity: Soft biopolitics and the modulation of control. *Theory, Culture and Society, 28*(6), 164–181.

Ettema, J., & Whitney, C. (1994). *Audience making: How the media create the audience*. Thousand Oaks, CA: Sage.

Fisher, E. (2015). You media: Audiencing as marketing in social media. *Media, Culture, and Society, 37*(1), 50–67.

Fisher, E. (2018). When information wanted to be free: Discursive bifurcation of information and the origins of Web 2.0. *Information Society, 34*(1), 40–48.

Flichy, P. (2007). *The internet imaginaire*. Cambridge, MA: MIT Press.

Foucault, M. (1994). *The order of things: An archeology of human sciences*. New York: Vintage Books.

Fuchs, C. (2012). Google capitalism. *TripleC, 10*(1), 42–48.

Gillespie, T. (2014). The relevance of algorithms. In Tarleton Gillespie, Pablo J. Boczkowski, and Kirsten A. Foot (Eds.), *Media Technologies: Essays on Communication, Materiality, and Society*, pp. 167–194.

Groden, C. (2015, July 9). This app can tell if you actually like your friends. Fortune.

Horkheimer, M. (2002). Traditional and critical theory. In *Critical theory: Selected essays*. New York: Continuum.

Katz, E. (2009). Why sociology abandoned communication. *The American Sociologist, 40*(3), 167–174.

Keeper, P. (n.d.). People keeper.

Kotliar, D. M. (2020). The return of the social: Algorithmic identity in an age of symbolic demise. *New Media and Society, 22*(7).

Lazarsfeld, P. F. (1941). Remarks on administrative and critical communications research. *Studies in Philosophy and Social Science, 9*, 2–16.

Lee, M. (2011). Google ads and the blindspot debate. *Media, Culture and Society, 33*(3), 433–447.

Litt, E., & Hargittai, E. (2016). The imagined audience on social network sites. *Social Media and Society, 2*, 1–12.

Lowrie, I. (2017). Algorithmic rationality: Epistemology and efficiency in the data sciences. *Big Data & Society, 4*(1), 205395171770092.

Lunt, P., & Livingstone, S. (1996). Rethinking the focus group in media and communications research. *Journal of Communication, 46*(2), 79–98.

Lupton, D. (2016). *The quantified self*. Malden, MA: Polity Press.

MacKenzie, D. A. (2007). *An engine, not a camera: How financial models shape markets.* Cambridge, MA: MIT Press.

Mackenzie, A. (2015). The production of prediction: What does machine learning want? *European Journal of Cultural Studies, 18*(4–5), 429–445.

Mackenzie, A. (2017). *Machine learners: Archaeology of a data practice.* Cambridge, MA: MIT Press.

Marwick, A. E., & Boyd, D. (2010). I tweet honestly, I tweet passionately: Twitter users, context collapse, and the imagined audience. *New Media and Society, 13*(1), 114–133.

Mayer-Schönber, V., & Cukier, K. (2013). *Big data: A revolution that will transform how we live, work, and think.* New York: Houghton Mifflin Harcourt.

McGuigan, L., & Manzerolle, V. (2014). *The audience commodity in a digital age: Revisiting critical theory of commercial media.* New York: Peter Lang Publishing.

Napoli, P. (2010). *Audience evolution: New technologies and the transformation of media audiences.* New York: Columbia University Press.

Neff, G., & Nafus, D. (2016). *The self-tracking.* Cambridge, MA: MIT Press.

Neyland, D. (2015). On organizing algorithms. *Theory, Culture and Society, 32*(1), 119–132.

Pasquale, F. (2015). *The black box society: The secret algorithmic that control money and information.* Cambridge: Harvard University Press.

Pool, I. de S., & Schulman, I. (1959). Newsmen's fantasies, audiences, and newswriting. *The Public Opinion Quarterly, 23*(2), 145–158.

Pooley, J., & Katz, E. (2008). Further notes on why American sociology abandoned mass communication research. *Journal of Communication, 58*, 767–786.

Rogers, R. (2009). *The end of the virtual: Digital methods.* Amsterdam: University of Amsterdam Press.

Rouvroy, A., & Stiegler, B. (2016). The digital regime of truth: From the algorithmic governmentality to a new rule of law. *Online Journal of Philosophy, 3*, 6–29.

Rudder, C. (2015). *Dataclysm: Love, sex, race, and identity; what our online lives tell us about our offline selves.* New York: Vintage.

Sadeh, T. (2015). From search to discovery. *Bibliothek Forschung Und Praxis, 39*(2).

Stevenson, N., Jackson, P., & Brooks, K. (2001). *Making sense of men's magazines.* Cambridge: Polity Press.

Turow, J. (2011). *The daily you: How the new advertising industry is defining your identity and your worth.* New Haven, CT: Yale University Press.

Turow, J., & Draper, N. (2014). Industry conceptions of audience in the digital space. *Cultural Studies, 28*(4), 643–656.

van Dijck, J. (2013). *The culture of connectivity: A critical history of social media.* Oxford: Oxford University Press.

van Dijck, J., Poell, T., & de Waal, M. (2018). *The platform society: Public values in a connective world.* Oxford: Oxford University Press.

Weber, M. (1921). *Economy and society: An outline of interpretive sociology.* Berkeley: University of California Press.

Weber, M. (1958). *From Max Weber: Essays in sociology.* Oxford: Oxford University Press.

Zuboff, S. (2015). Big other: Surveillance capitalism and the prospects of an information civilization. *Journal of Information Technology, 30*(1), 75–89.

Zuboff, S. (2020). *The age of surveillance capitalism: The fight for a human future at the new frontier of power.* New York: PublicAffairs.

3 Can algorithms tell us who we are?

I have no more made my booke, then my booke hath made me, A booke consubstantiall to his Author.

– Montaigne

Know thyself, or let algorithms do so

In 2016, Google produced a video for internal use, entitled *The Selfish Ledger*. The video was leaked in mid-2018 to *The Verge* (Savov, 2018), allowing us a glimpse into *The Political Unconscious* (Jameson, 1981) of Google and its algorithmic imaginary (Bucher, 2018; Flichy, 2007). The video – nine-minutes long, immaculately produced – narrates a comprehensive cosmology for the future of big data. Specifically, it is concerned with the question of how technological means and devices can be used to know and understand human beings, in order, ultimately, to be able to let them lead better lives. It imagines a reality where each individual is represented by a personal ledger where all digital footprints are registered, accumulated, sorted, and rendered into knowledge about the self through mathematical and technological devices: algorithms, machine learning, artificial intelligence, and so forth.

In this chapter, I would like to suggest that this video offers a new grand-narrative, which links *media, knowledge,* and the *self.* I want to take the video as an opportunity to delve on this narrative by comparing it with the discourse concerning another media that was central to the emergence of a new type of knowledge of the self in modernity: the diary. I want to argue that these two very similar media (materiality speaking), the ledger and the diary, offer a very different means of knowing the self, and indeed assume different ontologies of the self. Beginning in the 17th century, the personal diary, as media and as a literary genre, became a central technique of the self, key to the constitution of modern subjectivity. Like the ledger, the diary too, served

DOI: 10.4324/9781003196563-4

as a *material media* to discover knowledge about who one was, and formulate what she wanted to be and how to get there.

My intention is neither to refute the possibility of algorithms to know individuals through data, nor is it to critique the actual content of that knowledge. Rather, I want to focus on the *mode of knowing*, that is, epistemology, that underlies *The Selfish Ledger*, and understand it within the historical context of the link between media and knowledge about the self, formed during modernity. My argument is that even if the selfish ledger would be able to acquire the same knowledge about the self as the diary writer has about hers – that is, if the gap between objective and subjective knowledge were to diminish – these two epistemologies assume a radically different conception of the self, which bears on our understanding of what it means to be human in contemporary digital society. Diary keeping assumes a subject, which becomes aware of itself; it is a media form through which subjectivity could be developed. It facilitates critical knowledge, which involves reflection and a strive for subjectivity, as an expansion of the self's realm of freedom. In contrast, the selfish ledger seeks to create knowledge about the self by sidestepping these tenets of subjectivity.

I proceed by first describing the video, highlighting the underlying analytical link that it makes between the ledger, knowledge, and individual users. In the following sections I historicize this analytic framework by referring the metaphorical selfish ledger to the bookkeeping ledger, and to the personal diary. My point of departure is actually the great similarity between the ledger and the diary as epistemic media forms: both make a fundamental assumption about the link between, on the one hand, meticulously recording entries (or data) per an individual in real-time and over time, and, on the other hand, the emergence of new insights, a new knowledge about reality. But what interest me, of course, is to draw the differences between them, most crucially, the relegation of subjectivity from the process of self-knowledge in the ledger, which Google sees as a model for algorithmic knowledge.

The selfish ledger: Decoding the self

The Selfish Ledger (Google, 2016) is a video made by Nick Foster, head of design at X (formerly Google X) and a co-founder of the Near Future Laboratory, and David Murphy, a senior UX Engineer at Google. The video is narrated as a short essay, proposing a new way to think about the relations between human-beings and data. Let me describe the video in some detail and highlight key narratives and rhetorical devices, before laying bare its new model of the relations between self, knowledge and media.

The video starts with a panoramic shot of a night sky, with the earth's surface taking up a thin strip at the lower part of the screen, an image that conveys the wonder and beauty of the universe, but also reminds us of its enormity and the tiny place that human beings occupy in it. It is an apt introduction to a proposal that sidesteps humans as the protagonists of history, a common theme in post-humanist discourse. The video's representation of the endless universe cuts to the enclosed room of a scientist/explorer. Placed on his desk are various measurement tools, a pocket knife, and drawing pens. Throughout its duration, the video regularly shifts between, on the one hand, the desk on which pictures of the main protagonists of the story will be laid (representing "knowledge", or a platform where knowledge can be conceived), and the world (representing "reality"), on the other hand.

The story is narrated by a scientist/explorer who anchors the selfish ledger in the history of science. It begins by introducing Jean-Baptiste Lamarck's early 19th-century idea that an organism's unique experiences during its lifetime change its internal code, which is then passed down to successive generations; what he called "the adaptive force". While this idea was later refuted and superseded by Charles Darwin's notion of natural selection, as the video duly admits, it goes on to suggest that this Lamarckian theory is now "beginning to find new homes in unexpected places", that is, in digital environments. Indeed, a fundamental rhetorical device of the video is a metaphorical comparison between DNA (as the "code" of organisms) and data (as the "code" of human behavior), and so the narrative goes back and forth between recalling the established knowledge about the former, and the speculative knowledge about the yet-unfamiliar latter.[1]

Moving now to our digital civilization, the video suggests that the trail of big data left by digital media users is as rich and valid a description of who they are as genes are for organisms. "Our actions, decisions, preferences, movement, and relationships" are all monitored and stored in what is metaphorically referred to as a personal ledger. This ledger of our data, the video suggests, "may be considered a Lamarckian epigenome: a constantly evolving representation of who we are". In biological discourse, the epigenome consists of a record of chemical changes to the genes, which may be passed down to an individual's offspring. According to this imagery, Google's ledger, then, consists of a record of changed behavior resulting in changing data patterns.

Going back to the desk and to the history of science, the video now introduces the 20th-century work of Bill Hamilton, who first suggested the radical idea that "the driving force behind evolution was not the individual but the gene". Hamilton suggested that genes, rather than organisms, were the quintessential unit of analysis to decipher the living world. The survival of genes was dependent on whether an organism's behavior benefited them. His work was later adapted by Richard Dawkins who coined the metaphor of *The Selfish Gene* (Dawkins, 1976). Dawkins attributed the motive for survival to genes, thus seeing "the individual organism [as] a transient carrier, a survival machine for the gene". While genes obviously do not have a volition in the sense that we attribute to conscious organisms, they can nevertheless be thought of to be selfish in that they are geared toward securing their own survival, rather than the survival of the organism of which they are part.

Closing the introductory episode, the video shifts back to the digital world, recalling that "user-centered design principles have dominated the world of computing for many decades". We have indeed all gotten used to being personally identified by digital platforms and being served personalized results based on our data track-record. The video then introduces a radically novel idea:

> What if we looked at things a little differently? What if the ledger could be given a volitional purpose rather than simply acting as an historical reference? What if we thought of ourselves not as the owners of this information but as custodians, transient carriers, or caretakers?

The answer is explored in the next three episodes of the video.

The first episode summarizes what we have already come to know about surveillance capitalism and how data it rendered into knowledge by digital platforms (Andrejevic, 2012; Cohen, 2004; Fuchs, 2011). But it adds to it the idea of a *goal-oriented ledger driven by users*. Users may set "the volition for their ledger" to optimize their results. The visual example in the video is that of a user who chooses between three orientations for his ledger's volition: eat healthier, protect the environment, or support local business. "Over time", says the narrator, "by selecting these options, the users' behavior may be modified and the ledger moves closer to its target".

Whereas in this episode there is a strong sense of a *user controlling his ledger*, using it to assist him fulfill his wishes, the next episode takes a step further and undermines the very notion of a sovereign self.

Now the tables are turned: "As [...] the notion of a goal-driven ledger becomes more palatable, suggestions may be converted not by the user but by the ledger itself". The ledger, then, will assume agency, articulating a volition that will drive its actions. What might this volition be? The video suggests that this volition arises from the ledger's *will to know*. For example, when "the ledger is missing a key data source which it requires in order to better understand this user". In other words, the moment the ledger detects a lack in its knowledge of an individual, it seeks to fill this gap. This shift of agency from the user to the ledger is articulated also visually as the video moves from an image of a person dominating the scene and directing the ledger to images of a ledger (represented by a data center) and devoid of any human beings. It is now the ledger itself that gains control over the appropriate purpose for the knowledge it creates.

The video illustrates what the abstract notion of a ledger gaining volition might look like through a story. The protagonist of the story is not the user but rather the ledger, which is missing a data point that it finds important for its operation: the user's weight. To satisfy its own volition – its will to know – the ledger offers the user scales for purchase, so it can close the data gap. When such an effort proves futile and the data needed is still out of reach, the ledger goes a step further and designs a scale, which *does not yet exist* but might be appealing to the user. The ledger does that by pulling resources about the user's tastes and preferences. If it succeeds in luring the user into buying the scale, the product can then be custom-made by advanced production technologies, such as 3D printing. The scale would then be used by the user and the ledger would get a hold of the data it was missing.

The final episode of the video takes a quantum leap, highlighting the social and civilizational significance of the new epistemology brought about by the ledger. Going full circle to its inception, the video suggests we look at the ledger from a Lamarckian perspective. Seen this way,

> the codified experiences within the ledger become an accumulation of behavioral knowledge throughout the life of an individual. By thinking of user data as multi-generational it becomes possible for emerging users to benefit from the preceding generations behaviors and decisions.

Ledgers, then, continue to live after users' demise, comprising an ecosystem of ledgers:

As new users enter an ecosystem they begin to create their own trail of data. By comparing this emergent ledger with the mass of historical user data it becomes possible to make increasingly accurate predictions about decisions and future behaviors.

The promise encapsulated in the ledger is neither merely that of producing better custom-made products, nor just gaining more intimate knowledge about the nature of any given user ("Our ability to interpret user data, combined with the exponential growth in sensor-enabled-objects, will result in an increasingly detailed account of who we are as people"); instead, it is a promise to uncover the true nature of our humanity, as a species. Such knowledge will, in turn, contribute to new solutions to humanity's most persisting problems: "As cycles of collection and comparison extend, it may be possible to develop a species-level understanding of complex issues, such as depression, health, and poverty".

The coda of the last episode goes furthest in forecasting the transformative potential of the ledger:

> Just as the examination of protein structures paved the way to genetic sequencing the mass multi-generational examination of actions and results could introduce a model of behavioral sequencing.

It imagines the possibilities for behavioral engineering by means of the ledger:

> As gene sequencing yields a comprehensive map of human biology researchers are increasingly able to target parts of the sequence and modify them in order to achieve a desired result. As patterns begin to emerge in the behavioral sequences they too may be targeted. The ledger could be given a focus shifting it from a system which not only tracks our behavior but offers direction towards a desired result.

And note: here the "desired result" is no longer determined by a human subject but rather by a ledger.

The epistemology of the selfish ledger: A summary

The video, then, suggests a new epistemology for understanding human beings, an epistemology that draws on key intellectual sources of the modern scientific framework, but which travels into new terrains thanks to digital media. The epistemology can be summarized along six tenets:

1   ***Code/data***: Code is the commensurable element allowing the comparison of qualitatively different universes – biological and social. DNA, the code of genes, which determines organisms' behavior (their epigenetics) is compared with digital data. These simple codes – comprising four bases and a binary electric code, respectively – allow the emergence of complex languages (and universes). Hence, code/data gives us access to the essence of life, in both the natural and digital environments; it provides the most unobtrusive, immediate and objective representation of the essence of organisms and of the human self.

2   ***Media***: Each data point – each gene, or each entry in the selfish ledger – carries very little information, hence it is insignificant in and of itself. It is only through the amassment of entries over space (in the case of genes) or time (in the case of the ledger) that data can be rendered into knowledge. What enables this rendering is a media, a ledger, which registers and organizes data points.

3   ***A population of one***: Since so much data exist per each individual, their rendering into knowledge through statistical methods can be executed also at the level of the individual, without needing to collate individuals into groups.

4   ***Omnivorousness***: The ledger must contain as much data as possible and from as many sources as possible. This pertains both to the totality of data points to be collected ($N = $ All), and the variety of variables. Since the selfish ledger has no explicit *a priori* theory about humans, it deems all variables and all types of data relevant.

5   ***Productivity***: The ledger does not merely register but can be productive. If data is the code of human behavior, and if it could be decoded, it can also be rearranged in order to intervene in behavior and modify it.

6   ***Volition***: The last and most far-reaching tenet of the epistemology underlying the selfish ledger is the idea of *knowledge without a human subjectivity*. This is encapsulated in the metaphorical attribution of volition (or selfishness) to the ledger. The ledger has a will to know. It seeks always to get more data. And if relevant data is missing, the ledger can elicit an action with the user in order to obtain it.

Self, media, knowledge: A historical view

The Selfish Ledger can be thought of sociologically as a work of imagination, seeking to delineate possible futures. Being produced by Google for internal use, however, makes this imaginary highly performative, or at least geared toward making it a reality. As so much research had shown, technological imaginaries are aiming to be prescriptive and direct whole organizations toward a utopian horizon (Flichy, 2007; Turner, 2006). Yet, rather than thinking *with* the video about the future – whether it is likely and desirable, how it activates pertinent actors, and so forth – I wish to think *contra* the video by locating Google's imaginary of the selfish ledger within a history of its underlying epistemic trinity, consisting of self, media, and knowledge.

This turn to the past is first and foremost encouraged by the use of deeply traditional concepts and models in *The Selfish Ledger*, most central of which is arguably the metaphor of the ledger itself. Originally, a *ledger* refers to a scripture or service book, which is kept in church. The term stems from *legger* and *ligger*, lie and lay in current use, denoting a book laying regularly in one place. Unlike other "books", the ledger is characterized by its immobility and its identification with one entity; it is not a book to be printed, distributed, and circulated, but rather to be kept in a specific location (not unlike the selfish ledger which literally *lies* on Google's servers). Later, and more commonly, the term gave its meaning to a book of accounting, summarizing transactions of individual businessmen or firms, and kept on premises. While various media for registering economic transactions go back millennia ago, in its book form, the ledger gained prominence in the later middle-ages in Europe (Gleeson-White, 2013; Sangster, 2016; Soll, 2014).

In an inquiry into the origin of double-entry bookkeeping, Basil (1947) highlights the link between the ledger as a media form and a new way of knowing. Bookkeeping, he argues, is a "craft" and a "technique", based on the "discovery of new media" (p. 263). Before the invention of double-entry bookkeeping (in early 16th century), transactions were recorded on "mere scraps of paper", or as entries in diaries "where the settlement of debts was indicated by the effective though untidy method of deletion" (Basil, 1947, p. 264). The essence of the new method was organizing and processing data in a "systematized" (p. 264) manner that revealed new knowledge which was hitherto inaccessible. The ledger was a novel means to record business transactions, which allowed its users to create a new type of knowledge about their business by objectifying transactions and following them over time. This innovative method assumed that the analysis of accumulated data registered in real-time over time will lay bare a new type of knowledge, which had hitherto remained inaccessible.

The personal diary

At about the same period, the late middle-ages, and entering more prominently into modernity, the invention of another, very similar media was beginning to take shape: the personal diary. In juxtaposing the diary with the ledger (both the historical and the one imagined by Google), my goal is to highlight two epistemologies of the self that were created concurrently. By recalling this history, I hope to historicize the discourse on the selfish ledger and lay bare its conditions of possibility.

Coming of age in the 18th and 19th centuries, the personal diary became a centerpiece in the construction of a modern subjectivity, at the heart of which is the application of reason and critique to the understanding of world and self, which allowed the creation of a new kind of knowledge. Diaries were central media through which enlightened and free subjects could be constructed. They provided a space where one could write daily about her whereabouts, feelings, and thoughts. Over time and with rereading, disparate entries, events, and happenstances could be rendered into insights and narratives about the self, and allowed for the formation of subjectivity. It is in that context that the idea of "the self [as] both made and explored with words" emerges (Taylor, 1989, p. 183). Diaries were personal and private; one would write for oneself, or, in Habermas's formulation, one would make oneself public to oneself. By making the self public in a private sphere, the self also became an object for self-inspection and self-critique.

The *dia*ry (*jour*nal in French, *tage*buch in German) is of a temporal nature, as the term suggests. "A diary is a document in which the writer records his or her experiences, thoughts and feelings shortly after they happen, in discreet entries, often dated" (Heehs, 2013, p. 6). Unlike memoirs or autobiographies, they are not retrospective and do not have an explicit plot, "they are written from day to day" (Heehs, 2013, p. 6). It is only with the passage of time that these discreet entries are interpreted and gain new meanings. The diary, then, is not an *archive* that registers personal occurrences, but a *workshop* where one actively presents oneself to herself, and subjects this self to critique.

The history of subjectivity, then, is intertwined with technologies of the self: ways, methods, and techniques by which people construct their subjectivity as both autonomous and related to society at large. The diary, as a genre and media, was a tool of the first order for the construction of the modern self: "not just accounts of what happened but ways of molding the stuff of the past into models of what the writers wish to be" (Heehs, 2013, p. 6). Diaries were not merely descriptive, but constructive as well, representing a model of the self as "a product of thought and reflection" (Heehs, 2013, p. 2). Diaries, in the words of Susan Sontag (2009, p. vii), were "an exemplary instrument in the career of consciousness" (quoted in Heehs, 2013, p. 7).

Audience and subjectivity: Knowing the self by means of presenting

The aphorism "Know thyself", first attributed to the Oracle of Delphi, has accompanied Western thought since antiquity (Long, 2015). But it is only with modernity that this call for self-examination and self-knowledge was systemized and became integral to the notion of the self. By systematized, I mean that it became a robust social practice, embedded in a fuller cosmology, and interwoven with media. The assumption behind modern self-exploration, facilitated by epistemic media prone to reveal the self to the self, was "that we don't already know who we are" (Taylor, 1989, p. 178). The creation of knowledge about the self was geared toward creating identity, rather than revealing an already-existing one: "Rather than objectifying our own nature and hence classing it as irrelevant to our identity, it consists in exploring what we are in order to establish this identity" (Taylor, 1989, p. 178). As self-identity was sought after it was also formed.

Central to those experiments with the self was an "inward turn" (Taylor, 1989, p. 177), which contributed to a "radical reflexivity" (Taylor, 1989, p. 178). The very notion of the modern self, or subjectivity, arises from a new conception of interiority or inwardness, "the feeling that there is a personal inner space that we alone have access to" (Heehs, 2013, p. 3). While interiority and inwardness were conceptually crucial in opening up a space for self-reflection, the availability of a material space on which to arrange these reflections – a diary – was a no less important precondition: "The production of cheap paper [in the 16th century] had an enormous impact on the growth of the written word and writing had an enormous impact on the growth of the self-consciousness" (Heehs, 2013, p. 41). Writing, as Walter Ong taught us, separated the knower from the known, which, in turn, allowed for an "increasingly articulate introspectivity, opening the psyche as never before not only to the external objective world quite distinct from itself but also to the interior self against whom the objective world is set" (Ong, 2002, p. 103, quoted in Heehs, 2013, p. 41). Writing on paper created a space for the self to develop, a space where the self could be objectified in the world, thus creating subjectivity. We must therefore be constantly reminded of the triad self/media/knowledge and the different kinds of epistemology it facilitated.

Writing – to an actual interlocutor (in the form of letters), an imaginary interlocutor (in the form of letter novels), or to oneself (in the form of personal diaries) – thrived in 18th-century Europe: "Through letter writing the individual unfolded himself in his subjectivity" (Habermas, 1991, p. 48). Writing, says Habermas, was not merely an "imprint of the soul" and a "visit of the soul" but also a way to present the self to others, to make the self public. Likewise, "the diary became a letter addressed to the sender, and the first-person narrative became a conversation with one's self addressed to another person" (Habermas, 1991, p. 49). All these different media and techniques served what Habermas dubbed "experiments with ... subjectivity", centered on "audience-oriented privacy" (Habermas, 1991, p. 51). This was a radically new invention, linking self to knowledge through media. As Habermas puts it, the end of the 18th century "reveled and felt at ease in a terrain of subjectivity barely known at its beginning" (Habermas, 1991, p. 50).

Subjectivity, then, was formed by a dialectical force: It was at once "the innermost core of the private ... [yet] always already oriented to the audience" (Habermas, 1991, p. 49). Habermas refers to different media and spheres where publicity could take place, among which is the paper on which text could be written: newspapers, books, letters, and diaries. These media changed

> the relations between author, work, and public [...] They became intimate mutual relationships between privatized individuals who were psychologically interested in what was "human", in self-knowledge, and in empathy (Habermas, 1991, p. 50).

This has created what Habermas refers to as "directly or indirectly audience-oriented subjectivity" (Habermas, 1991, p. 49).

Subjectivity, then, is a specific form of self, which emerges with modernity and the Enlightenment. It emerges as a capacity of humans to *recognize* the limitations put on them by physical, natural, cultural, social, and political dictates, and seeking to *transcend* them by applying reason. What I am mostly concerned with here is the mobilization of reason to the task of freeing human beings not primarily from external constrains, but from constrains governing their thoughts, feelings, actions, and ultimately their sense of self, of who they are. Subjectivity, then, is a bridge toward an emancipated self. And media is central in facilitating the knowledge required for that task.

Protestantism as an impetus to know

As aforementioned, diaries are a modern phenomenon, appearing no earlier than the 16th century. They were used "for subjective expression as well as objective documentation" (Heehs, 2013, p. 8). But underlying documentation emerged a new practical layer, as diaries were used (similarly to the ledger) as a means of *improving* the self: "Along with self-expression came self-reflection, and along with self-reflection the desire for self-improvement" (Heehs, 2013, p. 8). An important impetus for this dynamic was the emergence of Protestantism, particularly Calvinism, and its doctrine of predestination. Why was self-examination so important if destiny is preordained? Without the ability to change their destiny, believers could at least hope to calm their anxiety by *knowing* their expected destiny; "They who know not themselves to be God's peculiar people will be tortured with continual anxiety" (Calving, quoted in Heehs, 2013, p. 48). Rather than encouraging fatalism, says Heehs, predestination actually drove believers "to inner examination" (Heehs, 2013, p. 47). Inner examination required a *technique*, a way to overcome the deceiving self, to lift the disguise over the truth that lies in the heart of believers: "Self-scrutiny was necessary because 'the heart of man has so many recesses of vanity, and so many retreats of falsehood, and is so enveloped with fraudulent hypocrisy, that it frequently deceives' itself" (Heehs, 2013, p. 47, quoting Calvin).[2] Self-examination required a more methodical way of knowing oneself, in order to overcome biases and self-deception.

The development of subjectivity, then, could be counted as another unintended consequence – to use Weber's formulation – of Protestantism, along with the rise of capitalism (Weber, 1958). The two developments are in fact entwined, making the link between the accounting ledger and the diary not merely homological but historical as well. While in the realm of the economy, Calvinists began keeping a close eye on business through meticulous bookkeeping, in their private lives, they began keeping personal diaries for much the same reason. "No man", says Calvin, "can arrive at the true knowledge of himself, without having first contemplated the divine character, and then descended to the consideration of his own" (quoted in Heehs, 2013, p. 65).

So indeed, "In the private pages of their diaries" just as in the pages of their business ledgers, "Puritans could confess their sins directly to God. The very act of writing granted them a sort of absolution" (2013: 49). As one notable diary writer explains, keeping a journal was meant to "give me a habit of application and improve me in expression; and knowing that I am to record my transactions will make me more careful to do well" (Heehs, 2013, p. 92). By the end of the 18th century, the practice of keeping diaries became widespread among "preachers, lawyers, merchants, soldiers, sailors, doctors, and a good number of writers" (Heehs, 2013, p. 91); "In New England ... 'almost every literate Puritan kept some sort of journal'" (Miller & Johnson, 1924, p. 461, cited in Taylor, 1989, p. 184). Diaries even "found [their] way into fiction" with notable literary protagonists such as Gulliver and Crusoe (Heehs, 2013, p. 91; Taylor, 1989, p. 184).

Between the ledger and the diary

The emergence of both the ledger and the diary stems from the will to know reality (be it economic or personal) while at the same time acknowledging the difficulty to obtain that knowledge. But despite similarities in the motivations that underlay their emergence, and in some of their forms, the resulting epistemology of the two media differ quite substantially. The ledger facilitated the *objectification of economic life*. What had been haphazard instances of commerce and exchange, embedded in personal relations and specific circumstances, became in the ledger a uniform entry of sums of money entering or leaving the business. The ledger has been a symptom of the rationalization of the economy, as well as a means of achieving that rationalization. The diary, on the contrary, and through quite similar media practices, facilitated the *subjectification of the self*. It rendered objective reality – of both self and the world – into a matter to be interpreted by the subject and molded into a sense of self. The diary has been a symptom of the rise of subjectivity in modernity, as well as a means of constructing subjectivity.

Underlying these particular historical events is a tripartite epistemic model of a self, created by a new way of *knowing*, facilitated by new *media*. Subjectivity is developed through mediated engagements, which allow the self to reflect upon itself, thus gaining new knowledge about itself. Self-exploration and self-reflection were central to the creation of subjectivity, since they articulated a new relationship between knowledge and the self. In the Enlightenment tradition, from Kant to Habermas, the construction of subjectivity expanded the scope of freedom for the self. That the subject is able to know itself is hence crucial for the resulting knowledge being critical, as it is knowledge that at one and the same time acknowledges the limits and biases of reason and insists on the ability of reason to overcome them.

The rise of subjectivity was not a result of the Enlightenment as much as it was a condition for its possibility. Core to the Enlightenment was the dictum to "act under the direction of reason rather than the coercion of authorities" (Heehs, 2013, p. 81). The use of reason for self-understanding and self-direction was central to Kant's definition of the Enlightenment as

> man's emergence from his self-imposed immaturity ... Immaturity is the inability to use one's understanding without guidance from another. This immaturity is self-imposed when its cause lies not in lack of understanding, but in lack of resolve and courage to use it without guidance from another. *Sapere Aude!* "Have courage to use your understanding!" – that is the motto of the enlightenment (Kant, 2009, quoted in Heehs, 2013, p. 81).

Conclusion: Knowledge without subjects

We are increasingly embedded within a socio-technical network prone on knowing us: individually, incessantly, in real-time, and geared toward shaping not merely our actions, but more fundamentally, our will and our sense of who we are. This endeavor of a socio-technical assemblage – a myriad of devices, technologies, and code, as well as bodies of knowledge, practitioners, industries, and discourse, which I call algorithms, as a shorthand – cannot be simply explained as an old practice done in a new way. Rather, this endeavor represents a new epistemology and a new ontology, a new definition of what it means "to know" human beings and understand them. To lay bare the assumptions underlying this new epistemology, I turned my gaze in this chapter to one of the single most important agents in articulating, defining, advocating, and working toward realizing it – Google. Using what can be termed a grounded-theory approach, I attempted to historicize the tripartite model of self, media, and knowledge by pondering the video's central metaphor, that of the ledger.

Juxtaposing the epistemology of the self that emerges from the narrative of *The Selfish Ledger* with the epistemology of the self that emerged during high modernity in the discourse on the personal diary, I sought to achieve two ends. First, to put the digital ledger on the same playing field as the diary, as far as the hermeneutics of the self is concerned. This helps us overcome a technologized, de-historicized, and fetishistic approach to studying algorithms. This I could do not only by highlighting the fact that the link between self, knowledge, and media is a longstanding one, rooted, in fact, in the emergence of modernity but also by showing that the ledger and the diary share many similarities in their approach to creating new knowledge by using new media: the recoding of simple events (or data) in real time, the accumulation of entries, their reinterpretation over time, and so forth.

Second, this comparison also highlights some dissimilarities, the most important of which concerns the engagement of *subjectivity* in the creation of knowledge about the self. I recalled the central role of the diary – a paradigmatic media of the Enlightenment – in forming a new kind of self in order to point out how algorithms are now facilitating the formation of another kind of self. What mostly characterizes the selfish ledger is the will to create knowledge, which is highly personal, data-driven, reactive, dynamic, but which dispenses with subjectivity. Not only does it attempt to bypass any subjectivity in the creation of knowledge about the self, but also its result is in effect the creation of a self, devoid of subjectivity, that is, an object (Rouvroy & Stiegler, 2016).

Subjectivity was not merely a precondition for the creation of knowledge about the self in modernity, it was also the product of such knowledge. What is perhaps most striking about *The Selfish ledger* is its fantasy of knowledge without subjectivity. But this is not a step back to the premodern, pre-Enlightenment epistemology which saw truth as independent of the knower, offering knowledge as myths, cosmologies, and creation stories; knowledge that is perceived as no different from reality. Rather, algorithms offer what might be termed a post-enlightenment epistemic model: quasi-scientific, data-based, machine-enhanced, behavioral, and decisively devoid of subjectivity and all that it entails – reflection, hermeneutical and critical capacities, and reason. As Antoinette Rouvroy eloquently puts it, "algorithmic governmentality", as she calls it, "does not allow for subjectivation processes, and this for recalcitrance, but rather bypass and avoids any encounter with human reflexive subjects. Algorithmic governmentality is without subject" (Rouvroy, 2013).

Notes

1 The unproblematic comparison between the natural world and digital technology, prevalent in the digital discourse, has been thoroughly explored. See, for example, Best and Kellner (2000) and Fisher (2010).

2 This notion holds in other sects as well: "The Puritan was encouraged to scrutinize his inner life continually, both to descry the signs of grace and election and to bring his thoughts and feelings into line with the grace-given dispositions of praise and gratitude to God" (Taylor, 1989, p. 184).

References

Andrejevic, M., (2012). Exploitation in the data mine. In Fuchs, K., Fuchs, K., Albrechtslund, A. & Sandoval, M. (Eds.), *Internet and Surveillance: The challenges of Web 2.0 and social media* (pp.71-88). London: Routledge.

Basil, Y. (1947). Notes on the origin of double-entry bookkeeping. *The Accounting Review, 22*(3), 263–272.

Beer, D. (2019). *The data gaze: Capitalism, power and perception.* London: Sage.

Best, S., & Kellner, D. (2000). Kevin Kelly's complexity theory: The politics and ideology of self-organising systems. *Democracy and Nature: The International Journal of Inclusive Democracy, 6*(3), 375–399.

Blasius, M. (1993). Introductory note to Michel Foucault's "about the beginning of the hermeneutics of the self: Two lectures at Dartmouth". *Political Theory, 21*(2), 198–200.

Bucher, T. (2018). *If… then: Algorithmic power and politics.* Oxford: Oxford University Press.

Cohen, N. S. (2004). The Valorization of Surveillance: Towards a Political Economy of Facebook. *Democratic Communiqué, 22,* 5–22.

Couldry, N., & Mejias, U. A. (2019). *The costs of connection: How data is colonizing human life and appropriating it for capitalism.* Stanford: Stanford University Press.

Dawkins, R. (1976). *The selfish gene.* Oxford: Oxford University Press.

Fisher, E. (2010). Media and New Capitalism in the Digital Age: The Spirit of Networks. New York: Palgrave.

Flichy, P. (2007). *The internet imaginaire.* Cambridge, MA: MIT Press.

Fuchs, C. (2011). The political economy of privacy on Facebook. Television and New Media, 13(2), 139–159.

Gleeson-White, J. (2013). *Double entry: How the merchants of Venice created modern finance.* New York: W.W. Norton & Company.

Google (2016). The Selfish Ledger [Video], Palo Alto.

Habermas, J. (1991). *The structural transformation of the public sphere: An inquiry into a category of bourgeois society.* Cambridge, MA: MIT Press.

Heehs, P. (2013). *Writing the self: Diaries, memoirs, and the history of the self.* London: Bloomsbury.

Jameson, F. (1981). *The political unconscious: Narrative as a socially symbolic act.* Ithaca, NY: Cornell University Press.

Kant, I. (2009). *An answer to the question: "What is enlightenment?"* London: Penguin Books.

Long, A. A. (2015). *Greek Models of Mind and Self.* Cambridge, MA: Harvard University Press.

Mackenzie, A. (2017). *Machine learners: Archaeology of a data practice.* Cambridge, MA: MIT Press.

Miller, P., & Johnson, T. H. (1924). *The puritans.* Cambridge, MA: Harvard University Press.

Ong, W. (2002). *Orality and literacy.* New York: Routledge.

Rouvroy, A. (2013). The end(s) of critique: Data behaviourism versus due process. In Hildebrandt, M. and de Vries, K. (Eds.), *Privacy due process and the computational turn: The philosophy of law meets the philosophy of technology* (pp. 143-168). London: Routledge.

Rouvroy, A., & Stiegler, B. (2016). The digital regime of truth: From the algorithmic governmentality to a new rule of law. *Online Journal of Philosophy, 3,* 6–29.

Sangster, A. (2016). The genesis of double entry bookkeeping. *Accounting Review, 91*(1), 299–315.

Savov, V. (2018). Google's selfish ledger is an unsettling vision of Silicon Valley social engineering. *The Verge,* May 17. Retrieved from https://www.theverge.com/2018/5/17/17344250/google-x-selfish-ledger-video-data-privacy

Soll, J. (2014). *The reckoning: Financial accountability and the rise and fall of nations.* New York: Basic Books.

Sontag, S. (Ed.). (2009). Writing itself: On Roland Barthes. In *A Barthes reader* (pp. vii–xxxi). New York: Barnes & Noble.

Taylor, C. (1989). *Sources of the self: The making of the modern identity.* Cambridge: Cambridge University Press.

Turner, F. (2006). *From Counterculture to Cyberculture: Stewart Brand, the Whole Earth Network, and the Rise of Digital Utopianism.* Chicago: University of Chicago Press.

Weber, M. (1958). *The protestant ethic and the spirit of capitalism.* New York: Scribners.

4 Can algorithms make aesthetic judgments?

With Norma Musih

Recommendation engines: Automating aesthetic judgment

Aesthetic judgment is a seemingly simple quest of valuating beauty in nature or in culture, of saying "this flower is beautiful", or "this is a good movie". It is a seemingly very personal and unreasoned decree; claims for beauty need not adhere to normative dictates, and they need no reasoning. Very much like love. Nevertheless, with modernity, a more complex view of aesthetic judgment emerged, epitomized in Immanuel Kant's *Critique of Judgment*. His view of aesthetic judgment was further developed, as well as politicized by Hannah Arendt. Following Kant, Arendt sees aesthetic judgment as both personal and social, requiring no excuses but begging for justifications, and as communal, communicative, and political. We wish to take this line of inquiry further by pondering what happens to aesthetic judgment in a culture populated by recommendation engines.

We therefore suggest the need to critically examine the ramifications of recommendation engines – as data-processing algorithms – neither just in terms of the biases they create (Crawford, 2016; Ferguson, 2017; Gillespie, 2012a, 2012b; Mayer-Schönber & Cukier, 2013) and which are hard to discern because of their opacity (Pasquale, 2015); nor merely because of their tendency to create a filter bubble (Pariser, 2012; Turow, 2011), or their inherent undermining of privacy (Dijck Van, 2014; Fuchs, 2011; Grosser, 2017; Hildebrandt, 2019; Kennedy & Moss, 2015). Rather we seek to highlight how recommendation engines change the very meaning of culture (Anderson, 2013; Bail, 2014; Gillespie, 2016; Hallinan & Striphas, 2014; Striphas, 2015). Furthermore, by doing so they also undermine our freedom by excluding a particular faculty of human subjectivity: making an aesthetic judgment. We seek, then, to understand recommendation engines as automating aesthetic judgments.

DOI: 10.4324/9781003196563-5

We proceed by introducing the central role of recommendation engines in contemporary culture. While corporate, professional, and popular discourse highlights the objective, data-driven, mathematical nature of algorithms, we hypothesize that underlying the technological work of recommendation engines are also ontological assumptions about the nature of aesthetic judgment. Based on an analysis of public discourse on recommendation engines in Amazon and Netflix, we discern two prominent ontological assumptions, asserting aesthetic judgment as objective and individualistic. In the following sections, we position algorithmic recommendation engines within the realm of the discourse on culture, rather than merely as technical devices, and offer a critique of their assumptions about aesthetic judgment by referring to Arendt's work. Such a discussion stresses the particularity of the cultural assumptions underlying recommendation engines, rather than their universality, and helps highlight the political implications of the algorithmic conception of culture.

Algorithms in culture: Between technical neutrality and political worldview

In recent years, culture has become increasingly mediated by algorithmic devices that organize, prioritize, and curate cultural content for users, based on data, derived from their interaction with digital platforms. Recommendation engines epitomize this algorithmization of culture (Carah & Angus, 2018), changing how individuals encounter cultural artifacts, such as books and films. That culture is mediated by cultural agents is not new. The cultural field has always been populated by multiple intermediators – for example, critics, gatekeepers, and curators – who engage in the delicate craft of highlighting cultural artifacts worthy of our attention. These cultural mediators recommend cultural artifacts based on their artistic quality, social significance, or relevance to readers. As culture increasingly takes place online, and as using it produces a plethora of data, digital platforms now render them into personalizes recommendation and become key intermediator, or "cultural *info*mediaries", as Morris aptly calls them, "increasingly responsible for shaping how audiences encounter and experience cultural content" (Morris, 2015). Recommendation engines have become consequential in determining our cultural intake in recent years, suggesting which videos to watch (YouTube), what songs to listen to (Spotify), and what posts to read (Facebook). The shift from established cultural intermediaries to algorithms introduces new logics to intermediation (Morris, 2015). We focus our empirical gaze on

Amazon and Netflix as they have become leading commercial distributors of cultural artifacts worldwide, and so their recommendation engines now play a central role in shaping culture.

In professional and popular discourse, the logic behind recommendation engines is seen as fairly transparent and straightforward. With access to a treasure trove of personal data about users' previous cultural choices, it seems plausible to be able to assess their cultural taste and make successful recommendations. It is indeed a maxim of *dataism* to consider data as representations of reality, and their mathematical manipulation as allowing the creation of objective knowledge (Dijck Van, 2014). However, a decade and a half of social research has taught us that algorithms are far from being abstract, technical, mathematical, and hence objective systems. Rather, they are imbued with social, ideological, and cultural presuppositions (Crawford, 2016; Gillespie, 2014, 2016; Pariser, 2012; Pasquale, 2015).

One way that social research on algorithms has sought to shed light on the politics of algorithmic systems is by attending to the ideological assumptions embedded in algorithms as an idea and a social process. Thomas and colleagues propose to examine algorithms as fetish – "social contracts in material form" – in order to unveil the "emerging distributions of power often too nascent, too slippery or too disconcerting to directly acknowledge" (Thomas et al., 2018). And Beer urges us to go beyond algorithms as technical and material in order to "explore how the notion or concept of the algorithm is also an important feature of their potential power" (Beer, 2017). We answer this call by asking how the discourse on recommendation engines is "evoked as a part of broader rationalities and ways of seeing the world" (Beer, 2017), or seeing culture, in our case. Recommendation engines should therefore be examined as an idea that promotes "certain visions of calculative objectivity and also in relation to the wider governmentalities that this concept might be used to open up" (Beer, 2017).

Following this approach, then, we ask "what kinds of politics do [algorithms] instantiate?" (Crawford, 2016). The ways by which recommendation engines picture the world through data represent, we argue, a particular worldview, which has ramifications for culture. We therefore reject the assumption that algorithms merely mathematically translate numeric data into knowledge, and are, therefore, indifferent to political, social, or normative concerns. In doing so, we join the longstanding social research into algorithms, which seeks to expose their worldview. Rather than suggesting that algorithms distort reality, as the notion of bias suggests (e.g., Noble, 2019), we use the notion of worldview, which requires no presuppositions about reality. We therefore see recommendation engines as operating based on a particular notion of truth (i.e., a discourse) about aesthetic judgment, which is not universal but particular and hence ideological.

Recommendation engines do not merely mediate culture but also change what culture means (Hallinan & Striphas, 2014). Being performative, they change the object they assume to measure – aesthetic judgment, in this case. Striking examples come from new data-based "aesthetic" categories created by Amazon, such as "Most-Wished-For books on Amazon.com", "Books Rated 4.8 Stars or Above", and "Page turners: books Kindle readers finish in three days or less". But the most paradigmatic of these new aesthetic categories is arguably "recommended for you", which offers personalized curation. These new categories of aesthetic judgment have stirred a vibrant public discussion. We unpack the public discussion about Amazon's and Netflix's recommendation engines through the analysis of articles published in major media outlets in the last decade.

Aesthetic judgment and recommendation engines

The link between culture and algorithms in general, and recommendation engines in particular, has become a topic of concern for public discourse in the last decade. At the most fundamental level, the rise of recommendation engines has meant that data has become central to the cultural field. The ability to automate and personalize recommendations requires a data-saturated media ecology. Making sense of data requires an abundance of data; the more varied the data are the better recommendation engines are able to identify and characterize users. This has made digital platforms "data-hungry". For Amazon, which first integrated data and algorithms into its book recommendations (Economist, 2019), this hunger for data affects its approach to books' retailing. When Amazon launched its own digital reader, *Fire*, in 2014, it saw it "less as a communication device than an ingenious shopping platform and a way of gathering data about people in order to make even more accurate product recommendations" (Economist, 2014). The Amazon reader, then, was less of a consumer product, and more a means of producing data in the "assembly line" of personalized recommendations.

Personalized recommendations, based on users' individual online behavior, account for 35% of Amazon's sales (Yek, 2017), and for 80% of the content viewed on Netflix (Chhabra, 2017). Figuring out users' tastes and likes – that is, predicting their aesthetic judgment – is a difficult task methodologically and technologically speaking. But perhaps more fundamentally, the question of what people want is of philosophical, psychological, and sociological nature. Algorithmically predicting aesthetic judgment, then, brings up not only methodological questions about *how* Amazon and Netflix know what we want, but also ontological questions about *what* it means to want. Put differently, the question is what aesthetic judgment entails in a digital culture. The discourse on recommendation engines helps us disclose the answer to this question. Our empirical inquiry reveals two dominant assumptions about aesthetic judgment underlying recommendation engines: an assumption concerning the *objective* nature of aesthetic judgment, and an assumption concerning its *individualist* nature.

I Aesthetic judgment as objective

One assumption underlying recommendation engines is that aesthetic judgment is an event that can be grasped objectively, with no recourse to subjectivity. The availability of quality, variety, and quantity of data allows an outside spectator (or a recommendation engine, in this case) to characterize with high confidence one's cultural taste, and predict their preferences. Quality, in this case, pertains to data as a good proxy for real-world behavior; variety and large quantity are needed in order detect patterned behavior in the absence of theoretical hypotheses about relations between variables. The idea of aesthetic judgment as objective is not new and is perhaps mostly epitomized by the work of Pierre Bourdieu. Bourdieu's theory of cultural taste (1979 [1984]) has sprouted a cottage industry of studies, which has proven the high correlation of cultural taste with socio-economic indicators. Commercial mass media have implemented his theory most prominently through the practice of segmentation – offering distinct cultural artifacts to distinct social categories, easily recognized and measured. Recommendation engines uphold this objectivist assumption, giving it a digital boost.

An early journalistic account of recommendation engines brings to the fore the question of aesthetic judgment as an objective reality to be discovered by algorithms. The title, "How they know what you like before you do" (Moser, 2006) evokes the enigma of a predictive technology, which excludes subjectivity from the process of judging. This ability of recommendation engines assumes that aesthetic judgment needs not involve a conscious act of free will. Recommendation engines are assumed to be devices that tap an objective and already-existing reality: taste. The notion of aesthetic judgment as an objective reality that can be gauged from data is articulated in Netflix's CEO vision, mentioned at the beginning of this book, that "one day ... we're able to show you exactly the right film or TV show for your mood" (Economist, 2019), a vision that assumes that judgment can be gauged without people's direct, conscious involvement.

The objectivist assumption to aesthetic judgment underlying recommendation engines is criticized by observers for ignoring the social coordinates of this technology. Zeynep Tufekci, a *Wired* columnist and media researcher, argues that algorithms make an aesthetic judgment *for* you, not *with* you, and are therefore promoting a mode of non-communicative knowledge. Recommendation engines, she says, reify culture, thus render aesthetic judgment into a derivative of an objective social structure. She points out a few concrete computational practices by which this takes place. One is making recommendations based on similar individuals:

> Behind every "people like you" recommendation is a computational method for distilling stereotypes through data. Even when these methods work, they can help entrench the stereotypes they're mobilizing. They might easily recommend books about coding to boys and books about fashion to girls (Tufekci, 2019).

Recommendation engines rely on prejudgments concerning what makes people alike. Such theory of taste – shared by both marketing professionals and reductionist applications of Bourdieu – renders subjective judgment redundant. It makes an air-tight correlation between social location and taste. While this can be shown to be objectively valid at a given point – that is, can be proven by quantitative empirical research – it also rules out the possibility for aesthetic judgment as an expression of subjectivity. This computational practice, then, objectifies taste and judgment.

Another computational practice of recommendation engines, creating yet another form of bias, concerns popularity. Algorithmic recommendations are influenced by identifying "trends" and prioritizing them. They "filter out common terms as background noise and highlight those that have acceleration and velocity on their side. This definition of trending buries ongoing conversations and amplifies sensational, new things" (Tufekci, 2019). Tufekci, then, reminds us that seemingly technical terms used to make recommendations – such as "people like you", or "trending" – are nonetheless socially constructed and hence political. More specifically, these seemingly neutral indicators of taste and judgment carry with them *a priori* assumptions about what taste and judgment are.

Some commentators find troubling the idea that recommendation engines seem to *reflect* who we are. This would suggest that recommendation engines create "a digital extension of ourselves" (Satola, 2018) over which we have no agency. With automated recommendations, it becomes harder to clearly demarcate the boundaries of a self, which has autonomous will and intentions: "Blurred lines now exist between our own original thoughts about what we might like and what an algorithm decides for us" (Satola, 2018). To the extent that the role of subjectivity in the process of aesthetic judgment is demoted, and to the extent that aesthetic judgment can be seen as an *effect* of objective causes, it is also more susceptible for external manipulations. Hence, recommendation engines can be seen as forming taste, not merely gauging it. An early journalistic account about the music service Pandora explains that it does not merely "connect listeners with all kinds of music"; rather, "the website's personalized music recommendations have sparked new listening habits" for users (Moser, 2006). The objectivist assumption means that the lines between *deciphering* a user's aesthetic judgment and *influencing* it become blurred.

Personalization entails not only which movies or books are recommended, but also how they are recommended. Users are assumed to be different not only in terms of their taste but also in how their taste can be solicited. Netflix, for example, may offer the same show differently to different users, that is, appealing to different aspects of their aesthetic judgment. Its algorithms personalize "how shows are presented to you" by using "different image tiles ... to entice different users" (Clarke, 2019). This procedure is done by creating objective aesthetic categories. A Netflix executive explains:

> "We break down a show into multiple themes, and then we create artwork to fall into all of those themes", she says. This means that each show has a number of potential tile images each user may be shown ... as people start to watch the content, Netflix's massive trove of data kicks in to inform who sees what (Clarke, 2019).

A tile may highlight the romantic plot for one user and the suspenseful plot for another, "or maybe all the tiles on your account will be of the female characters in a show, while [for others] flits between images of food and key props" (Clarke, 2019).

The objectification of aesthetic judgment by recommendation engines, with the resulting relegation of subjectivity, is seen by Amazon and Netflix as a solution to the problem of overabundance of choice, which cannot anymore be handled individually by users. "The beauty of having a virtual library is you have control and choice, but with a lot of choice, you can be overwhelmed", and so personalized recommendations are there to help you overcome this "choice paralysis" (Clarke, 2019). But relegating subjectivity from the process of judement is also perceived as problematic, even by Amazon and Netflix. Their recommendation engines need to strike a balance between determining users' choices and letting users retain a sense of control. The Netflix executive explains: "We're trying to navigate within that tension of making it easy and showing them the right information so they can understand what they want to watch, but not be overly invasive" (Clarke, 2019). Users' may be willing to get recommendations from algorithms, but their sense of autonomy should nevertheless be retained.

The relegation of subjectivity does not go unnoticed by users as well. "I look at my algorithm-generated 'Recommendations for Lizzie'" writes a columnist, "and I don't like that person – or the control involved in the process ... Netflix, in all its machine-learned wisdom, appears to know me better than myself" (O'shea, 2018). The columnist resists the idea that recommendation engines control her cultural horizons. Netflix may be right to think she likes romantic comedies, but she also does not want her cultural diet to consist only of them:

> Watching only my Netflix recommendations would be like using the internet only to look at cat pictures: reasonable on one level, but you would undeniably also miss out on some interesting stuff (O'shea, 2018).

The columnist concludes with a call to reassert subjectivity vis-à-vis algorithms: "Just as we should resist outsourcing our ethical decisions to machines, we should not allow them to make cultural ones for us either" (O'shea, 2018). This critique is echoed by another columnist who attempts to assert his subjectivity by manipulating the algorithm: choosing shows he does not actually watch "in hopes of having my preferences changed" (Beeber, 2019). A similar tone of critique of the objectification of taste is reiterated by *The Guardian* asking readers to report their "weirdest Netflix recommendations" (Lee, 2015), following a case where viewers looking for a film to watch in the vein of teen comedy *The Inbetweeners Movie*, reported Netflix suggested the holocaust drama *The Boy in the Striped Pyjamas*.

II Aesthetic judgment as individualistic

The second assumption concerning recommendation engines is that aesthetic judgment is individualistic. Underlying it is a model of culture as an assortment of artifacts from which individuals can pick. This individualistic assumption circumvents the role of inter-subjectivity in the formation of culture. While culture can be seen as individualistic, that is, as *dyadic* relations between individuals and culture, it can also be understood (as we will expand in the next section) as a social sphere constituted among individuals and based not only on consumption but on communication as well.

The inability of recommendation engines to access such communicative inter-subjectivity does not go unnoticed by observers, neither even by the digital platforms. An article in *The Atlantic* raises the conundrum why Amazon bought Goodreads – "a social network for book nerds with a devoted but far from enormous 16 million members" – for $150 million? The intuitive hypothesis would be that Amazon was after "a vast trove of data on Goodreads members" (Weissmann, 2013), data that would then be algorithmically analyzed in order to automate recommendations. This hypothesis is compatible with the objectivist assumption outlined earlier. The article, however, suggests a counter-hypothesis: it is the *failure* of algorithmic analysis of data, which led to the acquisition. Amazon noticed recommendation engines fail among avid readers; within this group, the power of recommendations lies still in personal interactions with other readers. Avid readers, 20% of the population who read 80% of books, now rely more on "personal recommendations from people they know" (Weissmann, 2013), received mostly through social media. This serves a blow to the algorithmic model: "What they're not relying on much more heavily are recommendation engines" (Weissmann, 2013). Amazon, then, acknowledges that recommendation engines ignore the communal and inter-subjective nature of culture, and therefore fail to produce good recommendations for avid readers. The article hypothesizes, then, that Amazon has bought a very old-fashioned technology of a vibrant literary universe in order to "transmit the recommendations of prolific readers to the average reader" (Weissmann, 2013). Since "11 percent of book buyers make about 46 percent of recommendations" (Weissmann, 2013), the cultural conversation that takes place on Goodreads is valuable for Amazon.

The imperfect ability of recommendation engines to gauge aesthetic judgment, mentioned at the end of the previous section, can also result from their neglect of the communal and discursive nature of culture.

While culture carries moral, normative, and political undertones, these are overlooked by algorithms; algorithms deal with data, which serve as proxy for culture, not with culture per se. An article in *Wired* points to the biases that this agnostic approach yields. "Curation algorithms", the article argues, "are largely amoral. They're engineered to show us things we are statistically likely to want to see, content that people similar to us have found engaging – even if it's stuff that's factually unreliable or potentially harmful" (Diresta, 2019). For example, anti-vaccine books have topped Amazon's Best Sellers in "categories ranging from Emergency Pediatrics to History of Medicine to Chemistry" (Diresta, 2019). The reason is inherent to the operation of algorithms, which the article laments: "recommendation algorithms can be gamed to make fringe ideas appear mainstream" (Diresta, 2019).

Recommendation engines carry an assumption of methodological and epistemological individualism (Hayek, 1942; Weber, 1978); they are geared toward producing knowledge about individuals and for individuals. Within this framework, "culture" is reified. One indication for that is the terminology of prediction, discovery, and serendipity, which prevails the discourse on recommendation engines (McCarthy, 2017). These terms reveal the presumed type of relationship between humans and machines underlying recommendation engines. *Prediction* assumes that given enough relevant data, recommendation engines can find out which cultural artifacts an individual may like to consume. Prediction is presumably descriptive, assuming to describe an event that will take place in the future with some probability. In the case of recommendation engines, however, prediction is also performative; its very existence is geared toward changing the probability of an event to occur. Amazon does not predict that someone may buy an item; instead it seeks to mobilize this "prediction" in order to increase the probability that she will. *Discovery* refers to a desired characteristic of recommendation engines – their ability to break the closed-circuit feedback loop, which would recommend users "more of the same" and create a filter bubble around them. To overcome this problem, recommendation engines strive to mimic the real-world experience of discovery, by programing *serendipity* into algorithms, which will allow users to happily run into new and surprising cultural artifacts as if they were in a second-hand indie bookshop.

Prediction, on the one hand, and discovery and serendipity, on the other hand, are quite contradictory, representing two poles of recommendation engines. Indeed, much of the discourse on recommendation engines, propagated by Netflix and Amazon, concerns the need to balance between giving people what they want (the prediction pole) and surprising them with new cultural artifacts (the discovery/serendipity pole). A Netflix executive explains:

> The average consumer is going to look at 40 to 50 titles to make their choice, so we have to put the right 40 to 50 titles in front of them without falling into the filter bubble. We have to make sure there's diversity and serendipity in those, and we have to use the signals of what they've told us before (Clarke, 2019).

A recurrent normative argument for the discovery potential of recommendation engines is the democratization of cultural taste. Recommendation engines expose users with low cultural capital to cultural artifacts they would not have accessed otherwise. This may be seen as a second coming of Walter Benjamin's insistence on the liberating potential of mechanical reproduction (Benjamin, 1969). This promise is critically explored in *The New Yorker* article on the fine-art department in Amazon:

> Amazon does not seem particularly interested in recommending art that subverts expectations or disturbs the comfortable. In fact, Amazon's model of personalized recommendation and mass appeal explicitly undermines the possibility of discovery that art dealers compete to offer. Instead, the store's "window displays" exhibit what its semi-automated gallerists think users want to see: famous names, bargain prices, and kitsch by the yard ... Discovery is left to the experts, and the hegemony of high culture, far from being undermined by Amazon, is reinforced (Mauk, 2013).

The article shoots three arrows of critique at recommendation engines, arguing, first, that they recommend art that's banal and comfortable rather than subversive and new; second, that they cannot compete with a human curator; and hence, third, that they do not actually democratize taste. Instead, recommendation engines flatter users, refrain from challenging them, and as a result fortify, rather than subvert traditionalist, consensual art. This image of culture as a stagnate, anti-communicative sphere, promoted by algorithms, is compared in the article with an idealized image of what the Web could have done for culture, and for discovery in culture:

Benjamin's arcades and their related concepts ... have been linked to the Internet before, with writers saying that in the early days of the Web one could idly wander, flâneur-like, through virtual spaces. But, just as arcades were replaced by the efficient shopping experience of the department store ... the Wild West of the early Web has been replaced by a thoroughly organized virtual space (Mauk, 2013).

The Web, which could have potentially led to more exploration and discovery in culture, became overly controlled by algorithms that undermine it as a communicative space.

This romantic-humanist critique of recommendation engines is re-iterated by many commentators, lamenting the narrowing-down of engagement with culture due to the individualization that recommendation engines promote. One article compares the experience of searching for books online with that of searching for them in a bookstore:

There is something special about walking into a bookstore and exploring the collection ... if it is my first time there, I tend to follow a similar path through the store to get myself acquainted. First, I float toward the literature section and walk across the wall from A to Z, scouring through the names of authors both familiar and unknown. Then, my eyes wander to the history and philosophy sections, where I can usually find esoteric titles that sometimes hint more at the tastes of the bookstore employees than the interests of their customers (Satola, 2018).

Such wild and unexpected discovery stands "in stark contrast to the clickbait world of the internet" (Satola, 2018). Amazon's recommendation algorithms "are destroying the humanistic side of reading and how we share books with others", that is, destroying the communality of culture. Whereas our stroll through a bookstore is described as a process of discovery, where we can at least get a glimpse of an actually existing "culture" as a social phenomenon, "the Amazon algorithm is set up so that you only see what the site wants you to see ... [which] reinforces your current tastes and opinions" (Satola, 2018). While the former expands your horizons and makes you face culture-at-large, the latter narrows it and blinds you from the social and communal character of culture: "What algorithms take away from the modern reading experience is its crucial interpersonal dimensions" (Satola, 2018).

A few commentators see in recommendation engines as offering a new means for making culture communal and inter-subjective, rather than individualistic. By connecting interrelating personal data from different users of a music service, algorithms are able to facilitate social communication among them, making "'music discovery' a social activity" (Moser, 2006). This, according to a study cited in the article, will lead to a democratization of taste:

> Instead of primarily disc jockeys and music videos shaping how we view music, we have a greater opportunity to hear from each other... These tools allow people to play a greater role in shaping culture, which, in turn, shapes themselves (Moser, 2006).

The study found that 58% of participants reported being exposed to "a wider variety of music since using any online music service" (Moser, 2006). This view upholds recommendation engines' role in revitalizing culture: "People are so hungry to get reconnected with [new] music" (Moser, 2006). Netflix's recommendation engine is also interpreted in the same light: the more than one billion ratings contributed by customers on its site (as early as 2006), rendered into algorithmic recommendations, accounts for 60% of the movies rented. The article refers to these algorithmic recommendations as "community-driven" (Moser, 2006), suggesting that it reflects a kind of social communication about culture.

But, are relations among data points really a form of social communication? An article delving on this question distinguishes between our true self (which underlies our taste) and our algorithmic self (which underlies how algorithms interpret our taste). Netflix, the article argues, "can find out what you like, but it can't read your mind" (Ditum, 2019). For algorithms, "there is no 'you'" (Ditum, 2019), only how you act online. The article distinguishes between what might be called an intersubjective assessment of someone's taste and an algorithmic assessment. A friend giving us an accurate recommendation for a book is a proof that she really knows us, since "in the usual version of ourselves, taste is at the center" (Ditum, 2019). With algorithmic judgment of taste, however, we enter a new realm where this very ontology of selfhood is denied:

> When it comes to Netflix, I simply don't exist. There's a general assumption that a service such as Netflix must be profiling you ... But that's not how Netflix works. All it knows is what you watch,

and what other people who watched those things also watched. Even the word "people" in that sentence is arguably out of place: there are no people in the Netflix algorithm, only relationships between shows and movies (Ditum, 2019).

This can be read as a humanist critique of a post-humanist cultural field, where mathematical connections between objects take over intersubjective relations mediated by natural language. The article exemplifies the difference between mathematical and natural language with the category of "race", integral to natural, intersubjective language in American society. When black Netflix viewers were served "thumbnails that highlighted black actors who were bit-players rather than the stars of a movie", they assumed the service knew their race and attempted to lure them. But that, according to the article, is not what happened. Netflix was serving them these thumbnails "not because it had any clue who the users were, only because it knew that they had previously watched shows advertised with thumbnails of black actors" (Ditum, 2019). Put differently, a user is not assumed to be black by the recommendation engine, but rather only to feature a particular data pattern (which in natural language we might describe as black).

The divergence of mathematical language from natural language is evident in how recommendation engines "talk" about culture. Cultural forms constructed by recommendation engines do not necessary coincide with those formed among people. A good example for that are genres. Artistic genres, such as tragedy, comedy, farce, or drama, are longstanding and, with some variations over time, still constitute a common language to discuss culture. A genre possesses its own formal structure, logic, and tropes. As aesthetic categories, genres allow communication among different actors; they allow making an otherwise idiosyncratic artwork a public matter. For algorithms, however, these socially and historically constructed categories are problematic as a basis for recommendations. Netflix finds genres to be

> too broad to help users find new content. Why settle for "drama" when you can have 'imaginative time travel movies from the 1980s' instead? Changing the way titles are categorized by becoming much more specific helps Netflix recommend quirky shows and movies that users may not find otherwise (Mcconnell, 2017).

Note the different conceptualization of genres assumed and promoted by recommendation engines. Rather than as *discursive* categories constructed in a public sphere, genres are perceived and acted upon by algorithms as *behavioral* categories. The more cultural taste is personalized by algorithms and the more it is objectified, so it is driven to be fragmented into micro-genres, which may be meaningless publicly. A Netflix executive explains:

> We can tell you how much violence or sex it has, does it have a dark ending or a happy ending … does it have a chimpanzee in it, does it have a corrupt cop, or does it have a corrupt cop who happens to be a chimpanzee (Mcconnell, 2017).

One would assume that not only are such aesthetic categories not universally discussed in culture but also that the user who supposedly made a favourable aesthetic judgment about them is not aware of them. Are they even categories of aesthetic judgment at all if the one supposedly making the judgment cannot vocalize them and advocate them?

Arendt's conception of aesthetic judgment and culture

The discourse on aesthetic judgment and culture, which underlie recommendation engines, is not universal but particular; as such, it differs from, what we term, a modernist view of culture. Where recommendation engines assume aesthetic judgment to be objective and individualistic, the modernist view sees aesthetic judgment as demanding the active participation of subjectivity and inter-subjectivity, and upholds culture as a communal sphere of communicating agents. We take Arendt's political rendition of Kant's *Critique of Judgment* (2012 [1790]), as an epitome of the modernist view. By engaging Arendt's view, we seek, first, to show that there is more than one conception of aesthetic judgment, and that the worldview underlying recommendation engines is therefore particular rather universal, and second, to identify the differences between these two views of aesthetic judgment in terms of their political ramifications.

Following Kant, Arendt asserts that aesthetic judgment is both social, since it relates to *sensus communis*, a 'community sense', and subjective, since there are no objective standards upon which it is based, that is, it does not refer to truth. This community sense is what makes humans capable of broadening their minds and thinking from the perspective of others (Degryse, 2011). Only when we are capable of thinking from other persons' standpoint are we able to communicate. We need a community in which our judgments and opinions can be vocalized and tested. Arendt therefore politicized aesthetic judgment: *Making aesthetic judgments and testing them in public condition and train our capacity for political judgment.* The validity of aesthetic judgment is anchored, according to Arendt, not in being objectivity truthful but in communicating subjective opinions, positioning one's judgment in relation to other people and to their common sense. Common sense "fits us into a community" (Arendt, 1992, p. 70); as it "makes us capable of thinking from the perspective of others ... it also enables us to speak to each other" (Degryse, 2011).

Community sense is "what judgment appeals to in everyone, and it is this possible appeal that gives judgments their special validity" (Arendt, 1992, p. 72). The validity of aesthetic judgment is anchored not in positivist truths but in the communication of subjective opinions in relation to other people and to their common sense. Aesthetic judgmeent, then, is subjective and inter-subjective, and the performance of both depends on their communicability. Appeal to community sense requires us to think, even subjectively, with a communal and communicative horizon in mind. It demands that we think as if we were a person among persons: "Only if we are capable of thinking from the other person's standpoint are we able to communicate. Without this capacity, we would not be capable of speaking in such a way that another person would understand us" (Arendt, 1992, p. 74).

Through speech and communication, common sense makes politics possible; moreover, it forms humans as political beings. Humans are connected to each other not only based on their needs and wants, as contract theories would have it. More important is our mental inter-dependence. Humans "are dependent on their fellow men not only because of their having a body and physical needs but precisely for their mental faculties" (ibid., p. 14); they are mentally interdependent and are bound to each other by a common world. Aesthetic judgment is, therefore, neither completely individual nor wholly social; rather, it is somehow both:

> One judges always as a member of a community, guided by one's community sense, one's *sensus communis*. But in the last analysis, one is a member of a world community by the sheer fact of being human; this is one's 'cosmopolitan existence'. When one judges and when one acts in political matters, one is supposed to take

one's bearings from the idea, not the actuality, of being a world citizen and, therefore, also a *Weltbetrachter*, a world spectator (Arendt, 1992, pp. 75–76).

When we judge, we always judge as members of a community, guided by what we all have in common.

Thinking subjectively with the *sensus communis* in mind, however, is not a substitute for taking into account the actual judgments of others, for having an actual dialogue with others (Arendt, 1992, pp. 43, 71). As Arendt puts it: "Unless you can somehow communicate and expose to the test of others … whatever you may have found out when you were alone, this faculty [i.e., enlarged thinking] exerted in solitude will disappear" (Arendt, 1992, p. 40). Communicating aesthetic judgment is, hence, part and parcel of making aesthetic judgment. "The very faculty of thinking", says Arendt, "depends on its public use; without 'the test of free and open examination', no thinking and no opinion-formation are possible. Reason is not made 'to isolate itself but to get into community with others'" (Arendt, 1992, p. 40). Degryse succinctly summarizes these two aspects of aesthetic judgment as a communal action:

> Our mental faculties call for others. Taking into account the possible judgments of others allows us to form judgments. But it does not stop here. We have to discuss our judgments and opinions with others in order to keep our mental faculties intact (Degryse, 2011).

Conclusion

Recommendation engines do not simply *automate* aesthetic judgment – as if leaving its essence intact – but rather change the action they set out to automate. This change is of both cultural and political significance. Culturally, recommendation engines presume aesthetic judgment to be objective and individual, thus undermining the subjective and inter-subjective character of culture. Our findings support existing research regarding the privatization and individualization of culture. Striphas warns that "what is at stake in algorithmic culture is the gradual abandonment of culture's publicness" (Striphas, 2015). In this ever-expanding private bubble "recommender algorithms … can act as 'intimate experts', accompanying users in their self-care practices", promoting, in turn, "creative self-transformation" (Karakayali et al., 2018). Just and Latzer (2016), who see recommendation engines as a new source of reality construction, evaluate that "compared to reality construction by traditional mass media, algorithmic reality construction tends to increase individualization" (Just & Latzer, 2016). And Prey underscores the performativity of recommendation engines, insisting that how they see the individual "work[s] to enact the individual on these platforms", and results in "algorithmic individuation" in the field of cultural consumption (Prey, 2018).

But we propose, following Arendt, that this change is also of political significance. According to the modernist view of culture, aesthetic judgment entails action – subjective and intersubjective. Recommendation engines relegate aesthetic judgment to the machine, disposing of the agential act of judging. Arendt sees not merely an analytical homology between aesthetic judgment and political judgment, but an empirical link as well: judging aesthetically in the cultural field also serves as training for political judgment. Recommendation engines render aesthetic judgments into choices, leaving out reflections upon these choices. Behavior is taken as a proxy for reflection, and choice replaces a deliberative process of argumentation and persuasion. As Arendt insists, what matters in aesthetic judgment is not so much what we choose, but that we are involved in the process of choosing as world spectators. That is what aesthetic judgment share with political judgment. By excluding this facet from the making of aesthetic judgment, recommendation engines focus on the end-product – the recommendation itself – the validity of which is assumed to appeal to notions of truth.

We have used Arendt's conception of aesthetic judgment as an epitome of a modernist ideal not in order to consecrate it against the algorithmic model, but rather for methodical and critical purposes. Methodically, this comparison compels us to understand recommendation engines as cultural intermediaries, rather than merely technical devices. Critically, this comparison, while not suggesting that one model of aesthetic judgment is more valid than the other, does help us to highlight the distinct political ramifications of each model. While the modernist model render culture a communal and communicative human endeavor, thus expanding its political horizons, the algorithmic model contracts these horizons.

References

Anderson, C. W. (2013). Towards a sociology of computational and algorithmic journalism. *New Media and Society*, *15*(7), 1005–1021.

Arendt, H. (1992). *Lectures on Kanta's Political Philosophy*. Chicago: University of Chicago Press.

Bail, C. A. (2014). The cultural environment: Measuring culture with big data. *Theory and Society*, *43*(3), 465–482.

Beeber, A. (2019). "Netflix Recommendations a Bit Unnerving." *The Lethbridge Herald*, February 2.

Beer, D. (2017). The social power of algorithms. *Information Communication and Society*, *20*(1). 1–13.

Benjamin, W. (1969). *Illuminations: Essays and reflections*. New York: Schocken Books.

Carah, N., & Angus, D. (2018). Algorithmic brand culture: Participatory labour, machine learning and branding on social media. *Media, Culture and Society*. 40(2), 178–194.

Chhabra, S. (2017). Netflix says 80 percent of watched content is based on algorithmic recommendations. *Mobilesyrup*. August 22.

Clarke, A. (2019). "How Netflix Decides What You Want to Watch." *The Sydney Morning Herald*, April 2.

Crawford, K. (2016). Can an. algorithm be agonistic? Ten scenes about living in calculated publics. *Science, Technology & Human Values*, *41*(1), 77–92.

Dijck Van, J. (2014). Datafication, dataism and dataveillance: Big data between scientific paradigm and ideology. *Surveillance and Society*, *12*(2), 197–208.

Ditum, S. (2019). *Netflix doesn't know who you really are*. *Australian Financial Review*, April 12.

Diresta, R. (2019). How Amazonas Algorithms Curated a Dystopian Bookstore. *Wired*, March 5.

Degryse, A. (2011). Sensus Communis as a Foundation for Men as Political Beings: Arendt's Reading of Kant's Critique of Judgment, *Philosophy & Social Criticism*, 37(3), (March 1, 2011): 345–58.

Economist. (2014). How far can amazon go? *The Economist*.

Economist. (2019, April 11). Amazon's empire rests on its low-key approach to AI. *The Economist*.

Ferguson, A. (2017). *The rise of big data policing: Surveillance, race, and the future of law enforcement*. New York, NY: New York University Press.

Fuchs, C. (2011). An alternative view of privacy on facebook. *Information*, *2*(1), 140–165.

Gillespie, T. (2012a). Can an algorithm be wrong? *Limn*, 2, 21–24.

Gillespie, T. (2012b). The relevance of algorithms. *Culture Digitally*.

Gillespie, T. (2014). The relevance of algorithms. In Tarleton Gillespie, Pablo J. Boczkowski, and Kirsten A. Foot (Eds.), *Media Technologies: Essays on Communication, Materiality, and Society*, pp. 167–194.

Gillespie, T. (2016). Trending is trending: When algorithms become culture. In Robert Seyfert and Jonathan Roberge (Eds.), *Algorithmic Cultures: Essays on Meaning, Performance and New Technologies*, pp. 52–75.

Grosser, B. (2017). Tracing you: How transparent surveillance reveals a desire for visibility. *Big Data and Society*, 4(1), 1–6.

Hallinan, B., & Striphas, T. (2014). Recommended for you: The Netflix prize and the production of algorithmic culture. *New Media and Society*, 18(1), 117–137.

Hayek, F. A. (1942). Scientism and the study of society. *Economica*, 9(35). 267–291.

Hildebrandt, M. (2019). Privacy as protection of the incomputable self: From agnostic to agonistic machine learning. *Theoretical Inquiries of Law*, 20(1), 83–121.

Just, N., & Latzer, M. (2016). Governance by algorithms: Reality construction by algorithmic selection on the internet. *Media, Culture & Society*, 39(2), 238–258.

Karakayali, N., Kostem, B., & Galip, I. (2018). Recommendation systems as technologies of the self: Algorithmic control and the formation of music taste. *Theory, Culture and Society*, 35(2), 3–24.

Kennedy, H., & Moss, G. (2015). Known or knowing publics? Social media data mining and the question of public agency. *Big Data & Society*, 2(2), 205395171561114.

Lee, B. (2015, July 10). What are your weirdest Netflix recommendations? *The Guardian*.

Mauk, B. (2013). The work of art in the age of Amazon. *The New Yorker*.

Mayer-Schönber, V., & Cukier, K. (2013). *Big data: A revolution that will transform how we live, work, and think*. New York: Houghton Mifflin Harcourt.

McCarthy, T. (2017, May 26). Amazon's first new york bookstore blends tradition with technology. *The Guardian*.

Mcconnell, J. (2017, September 8). Tracking TV habits; Netflix is refining recommendations for your next binge-worthy show. *Financial Post*.

Morris, J. W. (2015). Curation by code: Infomediaries and the data mining of taste. *European Journal of Cultural Studies*, 18(4–5), 446–463.

Moser, K. (2006, February 16). *How they know what you like before you do*. *Christian Science Monitor*.

Noble, S. U. (2019). *Algorithms of oppression: How search engines reinforce racism*. New York: NYU Press.

O'shea, L. (2018, April 16). What kind of a person does Netflix favourites think I am? *The Guardian*.

Pariser, E. (2012). *The filter bubble: How the new personalized web is changing what we read and how we think*. New York: Penguin Books.

Pasquale, F. (2015). *The black box society: The secret algorithmic that control money and information*. Cambridge: Harvard University Press.

Prey, R. (2018). Nothing personal: Algorithmic individuation on music streaming platforms. *Media, Culture and Society*. 40(7), 1086–1100.

Satola, A. (2018, September 17). A modern reading of humanism vs algorithm. *Michigan Daily*.

Striphas, T. (2015). Algorithmic culture. *European Journal of Cultural Studies*, *18*(4–5), 395–412.

Thomas, S. L., Nafus, D., & Sherman, J. (2018). Algorithms as fetish: Faith and possibility in algorithmic work. *Big Data and Society*. January 2018, doi: 10.1177/2053951717751552

Tufekci, Z. (2019, April 22). How recommendation algorithms run the world. *Wired*.

Turow, J. (2011). *The daily you: How the new advertising industry is defining your identity and your worth*. New Haven: Yale University Press.

Weber, M. (1978). *Economy and society*. Berkeley: University of California Press.

Weissmann, J. (2013, April 1). The simple reason why goodreads is so valuable to Amazon. *The Atlantic*.

Yek, J. (2017). *5 Lessons you can learn from Amazon's recommendation engine*. Disrupt by Altitude Labs. Retrieved from http://altitudelabs.com/blog/amazon-product-recommendation-engine/

5 Do algorithms have a right to the city?

Roads, traffic, and spatial politics

Governing traffic has always been a delicate practice, involving a network of drivers, devices, discourses, imaginaries, policy makers, city planners, laws, and regulations. While public attention has been given mainly to human and formal institutional actors, objects and technologies have also been central in shaping how people experience roads and in governing their movement (Fotsch, 2009). While it may seem up to individual drivers to decide which roads to use, their decisions are nevertheless caught within a web of actants enabling and limiting these choices – from roads and other vehicles to maps and road signs. In the past decade, a new actor came into the fray, a digital agent that participates in that assemblage, and that, to a large extent, threatens to override the governance of other actants: Waze. The navigational application acts as a decision-making actant, taking over much of the agency of drivers and other actants in figuring out routes to their destinations. But more than that, Waze offers a new way of looking at roads and traffic. In so doing, it also creates a new spatial epistemology, which is not always congruent with other spatial epistemologies and might even come into clash with them. Lastly, Waze offers knowledge, which is highly performative, affecting how space is known, experienced, and acted upon by other actants. It therefore also bares political ramifications, which are often contested by other actants.

While Waze promises to alleviate traffic, it also creates spatial effects considered to be problematic, such as the diversion of heavy traffic through side-roads, which have hitherto been outside of the radar of drivers, and seen little traffic. As Waze directs traffic through these small villages and serene neighborhoods, their (predominantly

DOI: 10.4324/9781003196563-6

affluent) residents contest the legitimacy of this new spatial practice. I use this novel spatial practice of algorithmic traffic diversion, and the contested discourse surrounding it, as vignettes through which to examine a new, algorithmic spatiality. I seek to, first, unravel the new spatial epistemology offered by Waze and informing individual traffic behavior; second, analyze how residents critique algorithmic spatiality and, third, how Waze responds to this critique and upholds algorithmic knowledge as a legitimate participant in shaping how space is used. I understand this struggle as a case of re-negotiating the "right to the city". Only this time, it is algorithms that raise such a claim. This claim, I further suggest, should be handled politically, rather than accepted as a technological necessity. Whereas the right to the city has traditionally been a critical theoretical and political means to assert the rights of underprivileged groups and individuals, it is currently upheld by an algorithmic assemblage, backed by some of the most powerful and opaque actors.

The chapter proceeds by outlining a theoretical framework weaving literature from diverse fields around the notion of algorithmic spatiality. The following, empirical, sections present residents' resistance to Waze's interference in space, followed by the discourse of other central protagonists: political authorities, critics, and most importantly, Waze. The chapter concludes by suggesting how the claim of Waze to offer a unique and superior knowledge about space is used to legitimate its engagement in space, or its "algorithmic right to the city".

Algorithmic spatiality

The promise of Waze – "outsmarting traffic together" – speaks to the power of digital technology to overcome old problems in new ways. In recent years, big data and algorithms have taken front stage in that dream of *The Digital Sublime* (Mosco, 2004; see also Mosco, 2014). Underlying it is the idea that by collecting real-time data from millions of drivers' mobile devices and processing them algorithmically, Waze is able to gain unique spatial knowledge and deliver personalized directions to drivers, thus streamlining traffic. To critically understand Waze, we first need to understand its epistemic stance, that is, the kind of knowledge that algorithms create about space. And second, since the knowledge that Waze creates has a large-scale impact on spatial reality, we need to lay bare its political ramifications, that is, the consequences of this knowledge in terms of social power (van der Graaf & Ballon, 2019).

Algorithmic spatiality represents a classic case of the power/ knowledge nexus, which is another way of saying that space is not objective but social and constructed, and as such, "spatiality is [...] a negotiated production" (Kitchin & Dodge, 2011, p. 74). A long lineage of critical geographers has theorized and empirically shown the subjective, inter-subjective, historical, social, and hence inherently contested and political nature of space (Eizenberg, 2013; Harvey, 2008; Lefebvre, 1992; Soja, 1989). Following Actor-Network Theory, we might more accurately think of space as techno-socially constructed, more so the more digital navigational media become dominant. Waze collects enormous quantities of user data, such as location, driving speed, place of origin and destination, and time of use, and renders them, with the aid of algorithms, into navigational knowledge. One productive way to theorize the new dyadic relations between algorithms and the city is Kitchin's notion of code/space (Kitchin & Dodge, 2011). Kitchin asserts a "mutual constitution" (Dodge & Kitchin, 2005) between code and space, "wherein how a space is produced, perceived and experienced is dependent on its mediation through code, and the spatial media is dependent on the encoding of spatial relations" (Kitchin et al., 2018, p. 12). The dyad suggests that there is no superior position to one over the other: we cannot say that code represents space any more than we can determine that space would be a simple reflection of code: "a code/space is dependent on the dyadic relationship between code and space. This relationship is so all embracing that if half of the dyad is put out of action, the intended code/space is not produced" (Kitchin & Dodge, 2011, p. 18). The enmeshments of these two very different ontologies is precisely what makes code/space a contested, political object, open for negotiation and change. The new constellation of code/space offered by Waze is a case in point.

Waze has become a central player in traffic management worldwide. While digital navigation devices have been in use at least since the 1990s, Waze is the first commercially successful system to integrate user-generated data, and provide personalized recommendations. It was founded in 2008 as a way to commercialize a community-project founded two years earlier: FreeMap Israel. By 2012, the company reported having 20 million users, and a year later this number reached 50 million. Waze was bought by Google in 2013 for almost $1 billion. At its peak, it was used worldwide by an estimated number of 130 million monthly users in over 50 languages. It serves here as an exemplar of code/space, juxtaposing algorithms, space, and society. Algorithms can be thought of primarily as epistemic objects, creating information and knowledge. But rather than being merely more sophisticated, advanced, efficient, objective, mathematical, and technological means to produce knowledge, they should be seen as creating a new type of knowledge, a new episteme. Moreover, their integration into mobile, personalized, and synchronic media technology renders their knowledge highly performative (Beer, 2009; Kitchin, 2017). The algorithms underlying Waze offer new knowledge about space, a new way of seeing space, while at the same time, changing how space is acted upon (Orlikowski & Scott, 2009). Or as Kitchin puts it, "Spatial media are more and more mediating how space is understood and the interactions occurring within them" (Kitchin et al., 2018, p. 12). We can therefore think about algorithmic spatiality as offering a new epistemology of space and a new politics thereof.

A central issue in the politics of space is "the right to the city" (Harvey, 2003, 2008; Lefebvre, 1992; Marcuse, 2009; Soja, 1989). Talking about "rights" in the context of a digital device deserves some clarification. The notion of "rights" is historically rich and politically complex. However, I use the notion of rights here in a very strict and contextual meaning, echoing its application in critical geography. In that spatial context, this term has been used as a critical-analytical tool to ask who gets a say on (mostly) urban matters when multiple actors are involved, some of which political (e.g., the municipality, residents), some exerting economic powers (developers, real-estate moguls), and some holding legitimate professional expertise (planners, architects). The thrust of this literature is concerned with analysing how different actors struggle to define, challenge, or fight to enjoy these rights. I continue this tradition and complicate the critical question by asking how a nonhuman actor, Waze, attempts to enter into the fray and negotiate its right to the city, and how other actors challenge that attempt.[1]

Upholding residents' privileged right to the city

The case-study on which this chapter reports – the public discourse on the diversion of traffic through side-roads in Israel – is a particularly fitting event to learn about the reproduction of code/space, as it represents a uniquely code/space phenomenon. Drivers may have always "cut through" side roads in the past, but only with Waze does this practice become a social problem and a site of conflict. As such, it helps us distinguish the different actors and their positions. The empirical corpus is based on news reports in local and national newspapers in Israel, as well as on reports and calls for action posted on social media by residents. All cases reported in this research reached the public eye because residents were actively protesting algorithmic traffic diversion.

Understandably, algorithmic traffic diversion is met with indignation and protest by affected local residents. Commonly, residents complain about the loss of their (privileged) right to the city and seek to reassert that right. This is most evident when the locale in question is privileged, either in terms of the socio-economic status of its residents, or the prestige it gained in the national spatial mythology. The villages of Kefar Netter and Beit Yehoshua at the center of Israel represent an intersection of these two privileged positions: both are weaved into Israel's national history and are inhabited by predominantly affluent residents. There are other locales reported here, like the village of Revaha, which represent a lower socio-economic profile. But all villages share a rhetoric of a closed, homogenous, and exclusive community, distinguished from its surrounding.

The thrust of residents' complaints is that high volumes of traffic diverted through their villages undermine their pastoral way of life. At the village of Revaha residents complain that

> our quality of life decreased substantially because of that. We are deprived of the experience and gains of rural life ... dozens of cars cut through the village's two lanes, and children are afraid to cross the roads as it became dangerous. We are residing on a thruway (Cohen, 2018)[2].

Residents of Revaha appealed to political authorities to counteract algorithmic traffic diversion. The District Authority (a regional government encompassing a few villages) gave residents permission to shut down the main entrance gate to the village, preventing nonresidents from passing through. The ordinance is constitutionally questionable, however. Many small villages in Israel are fenced, and are perceived (by both residents and nonresidents) as gated-communities, with the roads treated as private, although formally being public. Many villages commonly shut down their main gate after dark for safety reasons, practically preventing nonresidents from using the internal roads. Residents, then, simply used this means as a way to fight off traffic diversion by Waze.

The case is different in cities, where roads are more evidently public. There, residents resort to other means of resistance. In the affluent Green Neighborhood in the city of Kefar Saba, in Israel's central metropolitan area, residents complained that drivers shortcut through the neighborhood as an exit to Route 4, a major highway located just across the neighborhood. This created heavy traffic during rush hours (see Figure 5.1). Residents blamed Waze for turning their roads into highways, demanding the municipality to close-off this exit. On social media, neighborhood residents proposed to file a class-action lawsuit against Waze if it didn't refrain from diverting traffic into the

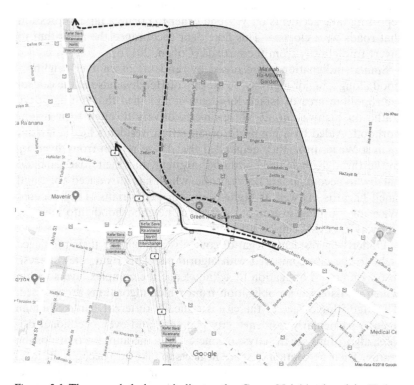

Figure 5.1 The rounded shape indicates the Green Neighborhood in Kefar Saba; the continuous line indicates the thruway from the city to Route 4; the dashed line indicates the diversion offered by Waze (Map source: Google Maps).

neighborhood (Tamarov, 2017b). A few days later, it was reported that the transportation committee of the municipality was due to discuss possible remedies to residents' complaints by implementing "traffic signs and enforcement" (Tamarov, 2017a). The municipality had already made similar promises in the past: in 2015, it announced that it was able to compel Waze to change its directions to drivers and prevent traffic from going through the neighborhood. The municipality insisted that it was able to pressure Waze to interfere with its automatic, algorithmic decision-making process and introduce man-made modifications. Waze denied that, stating: "we did not make changes in routes based on requests" (Keynan, 2015). Residents resorted to other tactics as well to resist their new spatial reality, such as approaching Waze with their grievances and demanding the company to overrule algorithmic agency with a human agency, or physically blocking roads. Another tactic involved hacking the algorithmic rationale itself by reporting fake accidents on Waze in order to create a false impression that roads were clogged. This fake data encouraged the algorithm to divert traffic away from these roads (Cohen, 2017b).

Similar indignations were raised by residents of another neighborhood along a major highway, in the city of Herzliya, also at the central metropolitan area of Israel. Residents complained that during heavy traffic on highway no. 2 Waze diverts drivers through their neighborhood. Aided by a nonprofit organization, residents filed a lawsuit against Waze. In it, they demanded that Waze refrain from diverting traffic through their neighborhood, arguing that their roads are too narrow to accommodate heavy traffic, which has introduced noise and smell hazards, as well as higher risks for pedestrians. They accused Waze of "turning overnight ... a quiet neighborhood into a major thruway" (Levi-Weinrib, 2016) (Figure 5.2).

Underlying residents' legal discourse is a claim for their privileged right to the city compared with algorithms. This right, they suggest, also assumes an obligation to acknowledge the complex layers within which roads exist, an obligation from which algorithms are excused. The algorithmic gaze of the city is radically different. Unlike human agents, particularly residents, city planners, and even occasional drivers, algorithms see the cityscape merely as a medium for transporting traffic. Their perception of space neglects to take into account the local texture of streets and neighborhoods. The thrust of residents' critique, then, is that algorithms ignore the complex social web within which roads are embedded, thus disrupt sustainable, humane spatiality. In

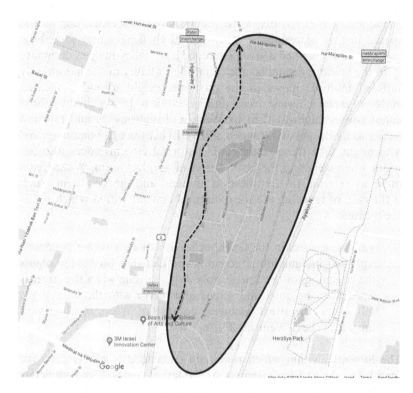

Figure 5.2 The circle indicates the Givat HaSofer neighborhood in Herzliya. Highway no. 2 is clearly marked, and next to it is the local route through which Waze diverts traffic (Map source: Google Maps).

more theoretical terms, we might say that algorithms superimpose an *abstract space*, where roads are thruways between starting and end points, over *lived space*, which is layered with interpretations, experiences, and practices (Lefebvre, 1992).

It is worth noticing the influence of contemporary digital discourse on the abstract space constructed by algorithms (Fisher, 2010). Space is imagined as a network, and cars as information; when a large quantity of cars needs to go from one node in the network to another, Waze breaks it down to packets and shoots it through different roads, or connectors. It is easy to see how in this framework roads are abstracted from their lived space. Indeed, in the lawsuit, residents accuse Waze for ignoring any spatial dimensions except traffic loads; it is indifferent to other spatial concerns, such as the spatial history of residents, other spatial challenges they face, how city planners imagined a given space, or how policy-makers designate it. In the lawsuit, residents argue that their quality of life has already suffered multiple blows in the past: the neighborhood is located in one of Israel's busiest areas, and has come to be engulfed by three major routes: highway 2 to its West, and highway 20 and railroad tracks to its East (Levi-Weinrib, 2016). In contrast to human agents, who might take these contextual, historical circumstances into account when weighing the justification of another harm, Waze treats those as external to the question at hand and ignores them. Such is the case of ignoring the local cultural context within which roads are located:

> synagogue goers in the neighborhood, as well as other residents, expressed indignation concerning the fact that on Friday nights [the beginning of the Jewish holy day] – a time when the street is expected to be quiet and serene and merge with the vibe of the holy day – the street becomes, because of Waze, bustling with noisy cars (Levi-Weinrib, 2016).

The lifeworld within which roads are embedded – in this case, the Jewish-Israeli, laid-back tempo of Friday evenings – is completely missing from the spatial gaze of algorithms.

But behind the claim that algorithmic spatiality ignores the specificity of space lies also a claim for privilege. The lawsuit asserts that "residents of what is supposed to be a quiet street in a quiet neighborhood, paying heavy local taxes, woke up one morning to find out that they are on the country's major highway" (Levi-Weinrib, 2016). While explicitly residents refer to the characteristics of the neighborhood ("quiet"), they also implicitly refer to their own socioeconomic characteristics. Simply put, residents uphold the unequally distributed privileges in urban space, which Waze ignores. Moreover, they rationalize their privilege as a right: the street and the neighborhood are quiet *because* they pay high taxes; the latter, then, is assumed as a condition for the former, and as its guarantor. The alleged blindness of algorithms to social privilege is seen as a threat to those who feel they hold a privileged right to the city.

"Undermining our way of life"

Residents at Revaha were also quite vocal in their protest against Waze. Speaking to a reporter, residents complained that the diversion of traffic through their village undermines their "way of life" (Ben-Zikri, 2018). The situation got so bad that during morning rush hour, cars had trouble getting out of their parking, and children and the elderly needed help to cross the road. Residents expressed their wish to return to a "sane level of traffic fitting a small village" (Ben-Zikri, 2018). A local official at Revaha puts the fault on Waze: "Waze doesn't give a damn about us... it's all their fault, they have no regards for the residents and couldn't care less. They put the residents of the village in jeopardy" (Ben-Zikri, 2018). As aforementioned, authorities closed off the village's main entrance gate, letting only local residents go through. This has led to quarrels with Waze users.

Another case of discontent from how algorithmic spatiality undermines residents' way of life arose in the Viznitz neighborhood of Bnei-Brak, an ultra-orthodox Jewish city at the metropolitan center of Israel. An ultra-orthodox online newspaper submits that Waze "might turn out to pose a real spiritual threat" to that community (Cohen, 2017a). The city, located near a major highway, has been suffering heavy traffic from "residents of neighboring secular cities" for years. It had tried in the past to prevent "secular" traffic from passing through the neighborhood by placing physical barriers at strategic points in the roads in order to discourage unwelcomed traffic. Now, however, with Waze, drivers find alternative routes and flood the neighborhood. This not only creates traffic jams – a common complaint in other cases – but also "threatens to undermine the Hasidic [pious] character of the neighborhood" (Cohen, 2017a).

Local religious leaders published a poster (a common means of communication in the ultra-orthodox community) calling residents to advise on how to tackle the problem and "preserve the purity of the Viznitz neighborhood ... from traffic coming from the secular city of Ramat Gan ... needless to mention the awful promiscuity of the drivers" (Cohen, 2017a). The poster further calls to "safeguard our eyes and the eyes of our children, lest we fail into witnessing, even by mistake, unwarranted sights" (Cohen, 2017a). Viznitz, then, is another case where algorithmic spatiality is seen as ignoring the embeddedness of roads in a specific culture. And here too, residents uphold their overriding right to the city as locals.

Political authorities, particularly municipalities, usually react quite vehemently toward Waze, seeing the app as undermining their sovereignty.

A case in point comes from the United States, but was widely reported in the Israeli press and served as precedent for local struggles. The municipality of Leonia – a small town in New Jersey, located at a strategic point near a major highway leading to Manhattan – took active measures to prevent nonresidents from going through town during rush hours, including issuing driving tags to residents and threatening non-compliant drivers with a \$200 fine. Officials acknowledged that these measures are extreme and constitutionally questionable, but justified them by the extremity of their predicament. Supporters of the quasi-closure of the town gave security concerns as justification for the measure: clear roads are needed, they assert, to ensure free access during emergencies (Be'eri, 2017).

The most elaborate struggle in Israel came in mid-2018, when disgruntled residents of Kefar Netter and Beit Yehoshua, two neighboring villages, frustrated by the impotent reaction of their respective municipalities, organized "action teams" to coordinate a grassroots disobedience movement to fight Waze. On June 4, they published a post on Facebook:

> Dear community, in light of the mayors' failure to respond to our appeal … in a letter sent to them three weeks ago, we have sent another letter on May 29 … as a last attempt to drive them to action. We … are also preparing for the possibility that this appeal too will not bear the desired outcomes and plan for various protest actions. We urge you to … join this important and necessary activity (Action, 2018).

The post features the two letters sent to authorities. The first letter, from May 8 – addressed to the mayors of the large neighboring city of Netanya, as well as other neighboring municipalities, and to the Minister of Transportation – is entitled "Solutions to transportation problems" (Committee, 2018b). In it, residents suggest building additional routes outside their villages as a way to solve the congestion problem in their villages. The letter ends on a rhetorical high-note:

> We are sick and tired of the negligence of our safety! Sick of the air pollution! Sick of the destruction of the rural character [of our village]! Sick of the ruin of our quality of life! Shan't we witness a total diversion of traffic of those who shortcut through our living spaces by May we will prevent cars from going

through, regardless of the consequences [of our action]. We hereby warn you – we will not allow that. We wholeheartedly hope that you will prevent unnecessary fights and act on behalf of your residents! (Committee, 2018b).

The second letter from May 29 was addressed solely to the Mayor of Netanya, entitled "A second and last warning concerning the detrimental effects of traffic hazards on the residents of Kefar Netter and Beit Yehoshua" (Committee, 2018a). The letter describes "in a nutshell … the outrageous, unbearable, dangerous and unreasonable situation" in which residents find themselves:

as the mayor of a city of tens of thousands of residents you have failed to provide infrastructures that can service thousands of cars a day(!) to highways and rail-stations, thus giving a fatal blow to two neighboring villages housing less than 300 households! (Committee, 2018a).

The letter continues outlining residents' plan for action:

We take public and moral responsibility to stop that negligence … We plan to use road blocks, aggressive protests, road signs, public relation [campaigns], and anything we see fit in order to completely sabotage the drivers that shortcut [through our villages], i.e., your [city's] residents! (Committee, 2018a).

Acknowledging that this is an aggressive move on their part, they nevertheless explain that they see no other choice but to sabotage the traffic routines of "those who chose to ruin our communities' life in order to shorten their route to work or to the train station by 10 minutes" (Committee, 2018a). The Facebook post also includes two posters to be printed and carried by residents during road blockades. The posters address residents of Netanya who cut through the villages. One of them reads: "Your municipality ignores (us) and you pay the price! Sorry, our roads are not meant for shortcuts". The other poster reads:

Dear short-cutter, we realize that you are in a hurry, and that in order to save a few minutes' ride, you couldn't care less about harming the life of others. We are determined to put an end to it! So before you ask yourself, tomorrow or the following day, why you cannot shortcut through our roads, and if you ignore our request, and try to make the shortcut anyway, you will find yourself at a standstill and arriving late at your destination (Action, 2018).

A few things can be learnt from these messages. Residents try to re-assert their right to the city vis-à-vis Waze through two means that counteract the rationale of power exerted by algorithmic spatiality. The first is to render spatial struggles *communal* rather than personal. Algorithmic spatiality makes knowledge about space and practice on space private and personal. Each driver receives instructions concerning how she or he should drive, implicitly ignoring their social-wide consequences. Both in the act of organizing as a collective, and in the messages directed to "short-cutters", residents reassert the communal and collective nature of space, making explicit the social-wide ramifications of personalized navigational recommendations. The second means by which residents try to counteract Waze is to render spatial struggles *physical* rather than virtual. While Waze presumably represents physical reality (roads, traffic conditions, etc.), it also constitutes a closed-off virtual space unto itself, where the real can be subsumed, hidden, or disappear altogether. By turning the struggle back to the physical world, residents tap into dimensions of space that Waze has almost no way to access.

Waze: Upholding the algorithmic right to the city

In all the cases discussed here, Waze is seen as the primary culprit. Hence, appeals by either residents or elected officials were made primarily to the company. Waze's standard reaction to these appeals is lean:

It's common knowledge that public roads are by construction open to everyone – regardless of whether they live in the neighborhood or not. Therefore, Waze directs drivers in all roads throughout the country, according to transportation laws and applicable traffic signs (Cohen, 2018).

Waze's main line of defense here is that the right to use roads is universal and egalitarian; local residents have no superior rights over local roads. Waze defines roads as ex-territorial: they are not (merely) an integral part of the neighborhood but also connectors between locales. Indeed, there is no such thing as *local* roads as they may not be local to all users.

In other public statements, Waze's rhetoric is more nuanced, allowing a vignette to a more layered view of how the company perceives algorithmic spatiality. In one statement, Waze acknowledges the superior rights of local stakeholders over roads, conceding that "our goal is to work holistically with our community of drivers, with maps editors, and municipal stakeholders, in order to improve driving experience for everyone" (Cohen, 2017b). This rhetoric of empathy and cooperation, however, is hindered by a discourse of legality and rights. Waze insists that it is subservient to universal legal and normative rules, rather than to any interest group. In the aforementioned case of Leonia, for example, where the local municipality closed-off 60 local streets from public use, Waze asserted that it would heed only if the streets were legally redefined as private. In the case of the Herzliya, neighborhood Waze further stated:

> given the goal of the service to provide direction suggestions based on the law and on terms of use, we are forbidden from preventing our users access to roads which are accessible and permitted for ride based on a request of that group or another (Levi-Weinrib, 2016).

Not only do algorithms have a right to suggest all roads to all drivers; moreover, by putting itself as subordinate and subservient to the law, Waze upholds its legal obligation to refrain from arbitrarily preventing drivers from using a road. This is indeed how residents, too, interpreted Waze's position, arguing in their lawsuit: "it is evident from Waze's reply that, unfortunately, it made up its mind to completely disregard the needs of residents, disregard the character of the street and the neighborhood through which it diverts traffic" (Levi-Weinrib, 2016). In response to the lawsuit by residents, Waze further highlights its formal legal obligating, stating that if residents wish to prevent the service from diverting traffic into their neighborhood, they should change traffic regulations, place stop signs and bumpers, and decrease speed limit as means of discouraging drivers from trespassing.

In response to a journalist's query, Waze offers an even more substantial justification for algorithms as stakeholders in space. The reporter raises residents' accusation that Waze does not relieve traffic jams but, in fact, creates them, to which Waze responds:

> public roads are intended for use by all citizens, among other things, in order to relieve traffic jams in busier areas of the city. We might use these streets within reason, but the algorithm of Waze will never create a traffic jam where it doesn't already exist (Mor, 2016).

This (indirect) exchange between residents and Waze seems to be a "charade" played by both parties, revealing a central tenet in the legitimation discourse of Waze for algorithmic spatiality. Residents complain about having traffic jams where once there were none – in their backyard. Waze responds to this accusation by delegitimizing the normative assumptions made by residents: the app, they insist, merely redistributes traffic jams more equally. By that Waze also underlines the illegitimate, hidden agenda of residents, who wish to exacerbate the unequal distribution of traffic jams by keeping them away from their backyard:

> Waze locates open road strips and distribute cars over a network of public roads, thus helping not merely alleviating traffic but also promoting safe driving, as crowded traffic often means greater risk for accidents and unsafe driving behavior. By diverting traffic from the most crowded bottlenecks in the city, Waze helps to reduce traffic jams across the city for all road users, as well as for local residents who don't use the application but often rip its benefits. The algorithm alternates among different roads, and it knows better than simply diverting everyone into a single road (Mor, 2016).

Waze, then, not merely subscribes to a legal discourse of rights but also puts forth normative and utilitarian arguments for its conduct. First, it argues that it helps create a more egalitarian space, where more roads, neighborhoods, and villages share the burden of high volumes of traffic. This makes algorithmic spatiality more egalitarian. Second, it argues that algorithmic spatiality makes road-use more efficient and convenient, perhaps not personally for each user, but commonly for all users. Waze deems the knowledge that algorithms produce socially beneficial, thus rendering them legitimate stakeholders in space, and upholding algorithms' right to the city. In reaction to similar complaints in the United States, Waze stated:

> [our] employees share the data they collect with urban planners, thus helping, for example, to monitor traffic light coordination [...] the company mobilizes a community of editors and volunteers that help insure that the maps are up to date and reflect municipal regulations. Our goal is to work together with the community of drivers, editors and city representatives in order to make everyone's driving experience better (Calcalist, 2017).

Waze's egalitarian ethos is upheld in an op-ed article by Na'ama Riba, a leading urban journalist and commentator. Riba criticizes the mounting indignation against Waze, coming primarily from residents of small, quiet villages. She categorizes residents' complaints about diverting traffic through their roads as NIMBY (Not In My Back Yard) struggles, where politically and socially powerful localities fight to prevent an unavoidable harm from hitting their space. "I am sorry to disappoint residents of suburbs and villages", writes Riba,

> but Israel of 2018 is a very crowded country. And I hate to break it to you – it is about to become even more crowded. Since no dramatic change in the public transportation system is in sight, in the next decade we are going to spend more time in traffic. And since the whole country is crowded, if traffic jams will be diverted from local roads in rural village, they will be added in a nearby town and will disturb someone else's pastoral life. And since roads in rural villages are public as well, there is no reason why they should not bare the national traffic burden. At any rate, Waze is certainly not at fault, as it helps to ameliorate the unbearable traffic, while making sensible use of public resources (Riba, 2018).

Not only are local roads a public utility, says Riba, they actually cost taxpayers up to four times as much as urban roads per resident. Furthermore, she puts some of the blame for heavy traffic on these very villages and their residents' way of life:

Crowded traffic in Israel is due not only to bad public transport, but also to low-density dwelling and to the fact that it has many small villages that require residents to own two cars per family – which leads to the construction of more roads (Riba, 2018).

Riba, then, rejects the technologically determinist position of residents, which puts the blame for heavy traffic on Waze. The reason for heavy traffic, she suggests, is the use of private, rather than public, modes of transportation; a problem exacerbated by people living in small villages. They are part of the cause for the problem, not only its victims.

Conclusion

This chapter examined the discourse through which algorithms legitimate their place in the production of space. It highlights two interrelated novelties emerging with algorithmic spatiality: epistemic and political. Epistemologically, algorithms "see" space differently and, hence, act upon it differently. They ignore the historical, cultural, and social contexts within which roads are embedded. This lack of embeddedness creates an image of space, which comes into clash with how other major actants, mostly human, perceive and experience space. Politically, algorithmic spatiality reshuffles the arena of power struggle among different actants over the right to shape space, where institutional and human actors now engage directly with a technological agent.

I used the notion of the right to the city as a vignette through which to examine how a nonhuman actant engages *politically* with human agents to curve out a place for itself in a socio-technical environment, and asserts itself as a legitimate bearer of rights. In doing so, Waze puts forth a few arguments, some of which are quite generic to digital platforms in general. But there is another argument that deserves our attention more carefully, precisely because it is founded on the novelty that algorithms bring about. According to it, not only is Waze not doing any harm, but it is actually making a significant social contribution because of its privileged access to knowledge, a knowledge, which is distributed, crowdsourced, egalitarian, unbiased, and non-political.

This epistemic claim allows Waze to assert itself as a legitimate political actor. Various actors claim that they have some unique and privileged spatial knowledge that other stakeholders do not, which legitimates their voice concerning space. Lefebvre's distinction of conceived, perceived, and lived space can be thought of along this epistemic axis, with each of the triad signifying a different form of knowledge, going from the universal, abstract, cognitive, and theoretical to the local, concrete, sensual and experiential (Lefebvre, 1992, pp. 38–39). The right to the city, then, is legitimatized, *inter alia*, by the quality and significance of the spatial knowledge that stakeholders possess.

Waze makes algorithmic knowledge part of the knowledge/power nexus that informs space, no less than that of city planners, residents, or political authorities. This is precisely the crux of the threat that Waze poses to others: questioning their epistemic privilege, and bringing into the fray a new spatial episteme. Algorithms challenge more established actors, who base their right to the city on privileged knowledge. Architects, city planners, urban scholars, and other spatial professionals and practitioners claim to be custodians of *expert knowledge* about space – scientific, objective, and rational. Political actors speak in the name of their privileged access to *democratic knowledge*, representing the will of the people. Capitalists uphold *market knowledge*, which signals what consumers want. And residents uphold their privileged access to phenomenological, *local knowledge*. Waze puts forth a claim for new knowledge to legitimate its right to the city. *Algorithmic knowledge* is unique, and no other actor has access to it. It combines the rational, objective knowledge of conceived space with the situated, mobile, fluctuating knowledge of perceived space, to use Lefebvre's terms. Waze knows in real time where people are, where they want to be, and how fast they're moving, and renders it into knowledge regarding the best way to use roads.

Algorithmic knowledge, according to Waze, is not only unique but superior to expert, democratic, market, and local knowledges, because it is more objective and egalitarian, devoid of interest and bias. It does not privilege one stakeholder over others, even challenging the very idea of locality, with its assumed privileges; as far as Waze is concerned, all roads are created equal. And so it claims to have unique ability to manage traffic in real-time in a way that leads to a more democratic and egalitarian use of space.

Algorithmic spatiality, then, represents not merely knowledge without subjectivity but something even more audacious: political engagement without subjectivity. From a normative standpoint, what stands at the heart of this story is not the question whether traffic should be diverted to quiet residential areas, and weather spatial privileges should be protected. Rather, what lies at the heart of the story is a question about governance: who gets to decide on privileges, rights, or any other political issue? And as this case study shows, slowly but surely, algorithms are increasingly threatening to usurp such decision-making powers from procedures, which involve subjectivity and inter-subjectivity. That to me, looks like an even bigger political question than the one about traffic redistribution.

Notes

1 This endeavor has been recently taken by a few scholars who engaged the idea of *The Informational Right to the City* (Shaw & Graham, 2017), or *The Right to the Smart City* (Cardullo et al., 2019).
2 Quotes from the corpus are translated from the Hebrew by the author.

References

Action, T. (2018). No Title.
Beer, D. (2009). Power through the algorithm? Participatory web cultures and the technological unconscious. *New Media and Society, 11*(6), 985–1002.
Be'eri, U. (2017). New Jersey launches a war against Waze. *Srugim*, December 27.
Ben-Zikri, A. (2018). Residents vs. Waze: A village in the south shut down its main road to drivers using the app. *Haaretz*, January 24.
Calcalist. (2017). The municipality that chose to fight Waze, which ruins residents' life. *Calcalist*, December 25.
Cardullo, P., Di Feliciantonio, C., & Kitchin, R. (Eds.). (2019). *The right to the smart city*. Bingley: Emerald Publishing.
Cohen, I. (2017a). Viznitz residents of Bnei-Brak vs. Waze: "Horrendous promiscuity." *Kikar HaShabat*, January 22.
Cohen, G. (2017b). No entry for Waze. *Channel 7*, December 26.
Cohen, S. (2018). Why does a whole village in the South boycott Waze? *Hidabrut*, January 25.
Committee, R. (2018a). *A second and last warning concerning the detrimental effects of traffic hazards on the residents of Kefar Netter and Beit Yehoshua.*
Committee, R. (2018b). *Solutions to transportation problems.*

Eizenberg, E. (2013). *From the Ground Up: Community Gardens in New York City and the Politics of Spatial Transformation.* London: Routledge.

Fisher, E. (2010). Contemporary technology discourse and the legitimation of capitalism. *European Journal of Social Theory, 13*(2), 229–252.

Fotsch, P. M. (2009). *Watching the traffic go by: Transportation and isolation in urban America.* Austin: University of Texas Press.

Harvey, D. (2003). The right to the city. *International Journal of Urban and Regional Research, 27*(4), 939–941.

Harvey, D. (2008). Space as a keyword. In Castree, N. Gregory, D. (Eds.), *David Harvey: A critical reader.* Oxford: Blackwell.

Keynan, E. (2015). Did Kefar Saba change lanes on Waze? *Ynet*, January 25.

Kitchin, R. (2017). Thinking critically about and researching algorithms. *Information, Communication & Society, 20*(1), 14–29.

Kitchin, R., & Dodge, M. (2011). *Code/space: Software and everyday life.* Cambridge, MA: MIT Press.

Kitchin, R., Lauriault, T. P., Wilson, M. W., Kitchin, R., Lauriault, T. P., & Wilson, M. W. (2018). *Understanding spatial media.* London: Sage.

Lefebvre, H. (1992). *The production of space.* Oxford: Blackwell.

Levi-Weinrib, E. (2016). Lawsuit: "Waze diverts traffic from Route 2 to a quiet street in Herzliya." *Globes*, December 1.

Marcuse, P. (2009). From critical urban theory to the right to the city. *City, 13*(2-3), 185–197.

Mor, Y. (2016). Waze to residents of Herzliya: Put stop signs in place. *Mako*, December 4.

Mosco, V. (2004). *The digital sublime: Myth, power, and cyberspace.* Cambridge, MA: MIT Press.

Mosco, V. (2014). *To the cloud, big data in a turbulent world.* London: Routledge.

Orlikowski, W. J., & Scott, S. V. (2009). The algorithm and the crowd: Considering the materiality of service innovation. *MIS Quarterly, 39*(1), 201–216.

Riba, N. (2018). The spoiled complaint of rural villages agaisnt Waze. *Haaretz*, February 4.

Shaw, J., & Graham, M. (2017). An informational right to the city? Code, content, control, and the urbanization of information. *Antipode, 49*(4), 907–927.

Soja, E. W. (1989). *Post modern geographies: The reassertion of space in critical social theory.* London: Verso.

Tamarov, M. (2017a). Following complaints: Transportaion committee to discuss limiting traffic in the Green Neighborhood during rush hour. *HaSharon Junction.*

Tamarov, M. (2017b). Resident of the Green Neighborhhod against Waze: "Drivers short cut to Route 4 through the neighborhood." *HaSharon Junction*, November 2.

van der Graaf, S., & Ballon, P. (2019). Navigating platform urbanism. *Technological Forecasting and Social Change, 142*, 364–372.

Epilogue

By the end of this book, I hope I have succeeded in realizing its two central goals. First, I have sought to offer a broad account of how algorithmic devices operate in society, particularly how they are being interwoven into our media-saturated ecology as knowledge devices. Here, my argument has been that algorithms offer a new way of knowing the world and the self, that they change the very foundations of what it means to know. This epistemic transformation also changes the objects on which knowledge is applied – culture, politics, society, and the self. My argument has been that algorithms and the knowledge they create cannot be solely understood in technical terms, but that underlying their operation are ontological conceptions concerning what human beings are.

Second, I have sought to put forth the argument that the new way of knowing that algorithms usher in – together with the ubiquity of digital platforms and devices geared toward monitoring users' data – is highly influential in social and political terms. Most prominently, I have suggested that algorithms seek to create knowledge which is increasingly independent of subjective and inter-subjective processes. It is founded on mathematical rather than natural language, it is performative rather than reflexive, and positivist rather than critical. It therefore excludes the central role that subjectivity has in the formation of knowledge about the world and about the self.

What ties the different chapters of this book together, then, is the idea that hitherto subjective and inter-subjective facets of social life are being partially taken over by an algorithmic episteme. It creates a new form of knowledge which can hardly be scrutinized by reason but which is rather fed back into our actions and choices, in effect, into how the (objective, subjective, and inter-subjective) world presents itself to us in digital environments. This undermines our subjectivity: our ability to think about our self for ourselves, to be self-reflective, to partake in creating knowledge about the self, to subject our thoughts, feelings, and wishes to a meta-process of critical thinking. It also undermines our intersubjectivity: the ability of independently reflexive subjects to converse and reach a mutual understanding (at the very least, an understanding of their disagreements) through natural language and through the process of communication.

The link between knowledge and subjectivity is historical and can be dated to the emergence of modernity and the Enlightenment. Subjectivity is the result of a new epistemology, a new way of knowing through critical self-reflection. But it is also the condition for creating this knowledge. The more the self is engaged in forming knowledge, the more it becomes a subject; and the more one becomes a subject the more one is able to form critical knowledge. As I have suggested, this link has been facilitated and become a dominant cultural form through the popular and quotidian use of media. How we think about our self, our conception of humans, is closely linked with the methodologies and techniques by which we come to learn about our self. And media has been central in the history of conceptualizing the individual for more than a century.

What this book has sought to do is to encourage us to think about how algorithmic knowledge changes our conception of the individual. Algorithms must not necessarily be seen as the cause for that transformation, but rather as the socio-technical expression of such as a move toward algorithmic governance, algorithmic self, algorithmic culture, and so forth. Algorithmic devices are an important site where this conception takes shape and is practiced. Vis-à-vis digital media, we learn how to dispense with our judgment and to put our trust in how they reflect back to us who we are.

The knowledge/subjectivity nexus, then, is historical, mediated, and dialectical. But it is also, and always has been, a political project in two senses of the term. First, in that it had a future-oriented teleology of emancipating individuals and enlarging their realm of freedom. And second, in that it did not assume subjectivity as an already-existing ontology anchored in cosmology, biology, or psychology, but rather a social construction, which needs to be maintained and legitimized, and which is always under threat of being deconstructed by counter-forces or toppled by alternative conceptions of the self.

Algorithm can be understood as the most recent threat to this humanist conception of the self. It is little surprise that the rise of digital media and the promise of artificial intelligence have been favorably accepted, if not outright celebrated, by post-humanist scholars. For them, the idea of an algorithmic self, a self increasingly molded by

algorithms – a hybrid human, codetermined by natural and technological processes – is seen as a promising route to overcome the
shortcomings of modernist subjectivity. Algorithmic knowledge about
the self is seen by them as liberating in that it undercuts the modernist
notion of the self as a coherent ontology and fosters a much more fluid
conception, unchaining individuals from rigid social categories. Post-
humanists celebrate the erosion of modern subjectivity by algorithms,
seeing that as opening up new horizons for the construction of a less
essentialized, more fluid and flexible self. The algorithmic self epito-
mizes for them the emergence of a techno-human cyborg, emancipated
from the metaphysical conceptions of modernity (Barron, 2003;
Braidotti, 2013, 2019; Brate, 2002; Fuller, 2012; Haraway, 1991, 2007;
Hayles, 1997a, 1997b; Shilling, 2005, p. 4).

The position put forth by this book is that the fantasy of a new
algorithmic self, who's identification is based on data patterns, and
discovered by algorithmic processing of data, rather than on natural
language, is politically dangerous. As Sherrie Ortner has argued, the
question of subjectivity is important because subjectivity is the basis
of political agency. Subjectivity is not merely individual; it is also a
structure of feelings, historically constructed and socially imposed, it is
part of public culture, of what connects individual actors into a collective
consciousness (Ortner, 2005). Subjectivity animates human action. The
question of subjectivity, as it was defined at least in the French-German
tradition, is a question of the dialectics of freedom and domination.
To ask about subjectivity and its conditions of possibility, as I have
tried to do in this book, is to ask to which extent subjectivity is free and
autonomous or a mechanism for domination? (Rebughini, 2014).

Making subjectivity redundant, as an ideal if not in practice, is
therefore politically dangerous. Our conception of self and identity, in
the sense of how individuals perceive and identify themselves, has
come to be based during modernity on ascription to social categories, a
belonging to groups of individuals who are identical. Self-identifying
as "worker", "gay", "African-American woman", and so forth, meant
not a complete assimilation or de-individuation. It did not imply that
all who belong to that group are identical in every way, but rather that
anyone belonging to a social category perceives itself as identical in
aspects that are politically significant, for example, suffering from similar forms of discrimination, or sharing the same economic interests.
Since their similarity to others in the group was understood in political
terms, their subjectivity was political as well. In a liberal society,
subjectivity has been the springboard for forming political claims and
fighting for political rights.

The mass media has been a trusted ally for this political conception of subjectivity, accompanying it, facilitating it, perhaps even helping to form it. Indeed, it is commonly criticized today for imposing a too totalistic and homogenous conception of individuals. It is in light of this critique that personalization, propagated by digital media, is universally celebrated for overcoming. But rather than merely celebrating the personalization of media we should also consider its deeper social and political significance. With all its problems, the mass media reflected and reproduced categories of identity that could be spoken of with natural language, understood theoretically, be subjected to critique, and even resisted through political action. The move to personalization in digital media was not merely a quantitative shift – collecting and analyzing more personal data – but a quantum leap involving a new epistemology. It involved a transformation in how individuals are seen.

The move to personalized media entails grouping individuals based on similar data patterns, rather than a similar identity. With its reliance on algorithms in order to characterize users, digital media categorize individuals based on data patterns, which cannot be understood with natural language, cannot be spoken about, and cannot be critiqued. Under such conditions, the very ontology of self is transformed: from a subjective construction of an individual who identifies himself or herself, and an inter-subjective construction of a multitude of individuals who identify each other through an interpretive and critical process set on a political horizon, toward an objective construction of the self, the meanings of which lay outside of the hermeneutic capacity of individuals. Algorithms, in short, contribute to the construction of a post-ascriptive, post-demographic self that undermines political subjectivity.

The algorithmic episteme undercuts the critical faculties inherent in narrative, speech, and inter-subjectivity. Critical knowledge and self-reflection are crucial to transforming individuals from objects to subjects. Hence, the creation of knowledge about the social world and about the self has to have a public and participatory component in order to incorporate subjectivity. In that sense, even if recommendation engines are able to characterize our taste and correctly predict what we would like to watch before we even think about that – and it is very likely they can, at least to some degree, as our taste *can* indeed also be thought of in objective terms – they already undermine the idea of taste as a project of subjectification, and the cognitive practices required to construct it. Under such conditions, even interpellation is undermined, and with it the possibility to be critical of it and resist it.

* * *

This book makes no attempt of giving a holistic picture of our algorithmic society. It merely presents the thrust of algorithms toward a particular direction. It tries to lay out what we are dealing with when we talk about algorithmic devices, and what their social ramifications may be. But this should not be seen as the end of the story of algorithms; we are very much still in its midst. The story of algorithms is also one of struggles, counter-forces, setbacks, and reorientations. Serious intellectual, ethical, and practical attempts are being made in the past few years to tackle the social havocs that algorithms are threatening to wrack. A number of popular books have been published in an attempt to refute algorithms' claim for creating a more objective knowledge (Christian, 2020; Criado-Perez, 2021; Eubanks, 2018; Noble, 2019; O'Neil, 2017; Wachter-Boettcher, 2018). Indeed, underlying these works is the notion that algorithms are not delivering on their promise: rather than overcoming problematic manifestations of subjectivity and inter-subjectivity, such as racism, they actually reify them. Scholars have therefore suggested the need to make the operation of algorithms more ethical by means of transparency and auditing (Brown et al., 2021; Mittelstadt et al., 2016; Guszcza et al., 2018; Kearns & Roth, 2019; Kim, 2017; Koshiyama, 2021) or demanding the creation of agonistic algorithms (Crawford, 2016; Hildebrandt, 2019). Such practices would help open up the black box of algorithms, make them more susceptible to critique, and allow more democratic oversight over what are mostly proprietary technologies.

Another attempt to counter the erosive effects of algorithms has been to educate users, making them more aware of how data is translated into knowledge (Carmi et al., 2020; Carrington, 2018; Gray et al., 2018; Pangrazio & Selwyn, 2019; Swart, 2021). Data and algorithmic literacies are supposed to help individuals to handle the algorithmic knowledge they regularly face more critically. In a similar vein, research has called our attention to the interpretive work of digital media users, and the efforts they make to resist algorithms' subversive effect of their agency and subjectivity. As we have seen in chapter 4, Netflix may try to guess your taste and make your judgment redundant but individuals may also resist that by trying to game the system, for example, by choosing particular films (without even watching them) that they assume would eschew the future choices of the recommendation engine in the desired direction. Likewise, in chapter 5, we have seen how local residents try to resist Waze's threat to their political sovereignty by reporting falsely about accidents, thereby using the very algorithmic logic of the application in order to divert traffic away from their neighborhood. There have also been many attempts – some have already come into fruition – to put forth policies and legislations to reduce the threat to privacy that algorithms entail, such as the European Union's General Data Protection Regulation (European Union, 2018). Such policies limit how much data can be extracted from users, how freely data can be shared across platforms, for which purposes data can be used, and so forth (Andrew & Baker, 2021; Hoofnagle et al., 2019). As we are heading toward a future of social life increasingly taking place over digital platforms, with more ubiquitous algorithmic devices analyzing more users' data, it is incumbent upon us to also question to what extent our subjectivity is made redundant by algorithms, to what extent is that a threat, and what we can do to counteract it.

References

Andrew, J., & Baker, M. (2021). The general data protection regulation in the age of surveillance capitalism. *Journal of Business Ethics, 168*(3), 565–578.

Barron, C. (2003). A strong distinction between humans and non-humans is no longer required for research purposes: A debate between bruno latour and steve fuller. *History of the Human Sciences, 16*(2), 77–99.

Braidotti, R. (2013). *The posthuman.* Cambridge: Polity.

Braidotti, R. (2019). *Posthuman knowledge.* Cambridge: Polity.

Brate, A. (2002). *Technomanifestos: Visions from the Information Revolutionaries.* New York: Texere.

Brown, S., Davidovic, J., & Hasan, A. (2021). The algorithm audit: Scoring the algorithms that score us. *Big Data and Society, 8*(1).

Carmi, E., Yates, S. J., Lockley, E., & Pawluczuk, A. (2020). Data citizenship: Rethinking data literacy in the age of disinformation, misinformation, and malinformation. *Internet Policy Review, 9*(2).

Carrington, V. (2018). The changing landscape of literacies: Big data and algorithms. *Digital Culture and Education, 10*(1), 67–76.

Christian, B. (2020). *The alignment problem: Machine learning and human values.* New York: W. W. Norton and Company.

Crawford, K. (2016). Can an algorithm be agonistic? Ten scenes about living in calculated publics. *Science, Technology & Human Values, 41*(1), 77–92.

Criado-Perez, C. (2021). *Invisible women: Data bias in a world designed for men.* New York: Abrams Press.

Eubanks, V. (2018). *Automating inequality: How high-tech tools profile, police, and punish the poor.* New York: St. Martin's Press.

European Union. (2018). *General data protection regulation (GDPR) – official legal text.* European Union.

Fuller, S. (2012). *Preparing for life in humanity 2.0.* New York: Palgrave.

Gray, J., Gerlitz, C., & Bounegru, L. (2018). Data infrastructure literacy. *Big Data and Society, 5*(2).

Guszcza, J., Rahwan, I., Bible, W., Cebrian, M., & Katyal, V. (2018). Why we need to audit algorithms. *Harvard Business Review, 4.* November 28.

Haraway, D. (1991). *Simians, cyborgs, and women: The reinvention of nature.* New York: Routledge.

Haraway, D. (2007). A manifesto for cyborg. In *Reading digital culture* (pp. 27–28). Oxford: Blackwell.

Hayles, K. (1997a). *How we became posthuman: Virtual bodies, cybernetics, literature, and informatics.* Chicago: University of Chicago Press.

Hayles, K. (1997b). The posthuman body: Inscription and incorporation in galatea 2.2 and snow crash. *Configurations, 5*(2), 241–266.

Hildebrandt, M. (2019). Privacy as protection of the incomputable self: From agnostic to agonistic machine learning. *Theoretical Inquiries of Law, 20*(1), 83–121.

Hoofnagle, C. J., Sloot, B. V. D., & Borgesius, F. Z. (2019). The European Union general data protection regulation: What it is and what it means. *Information and Communications Technology Law, 28*(1). 65–98.

Kearns, M., & Roth, A. (2019). *The ethical algorithm: The science of socially aware algorithm design.* Oxford: Oxford University Press.

Kim, P. (2017). Auditing algorithms for discrimination. *University of Pennsylvania Law Review Online, 166*(1), 10.

Koshiyama, A., Kazim, E., Treleaven, P., Rai, P., Szpruch, L., Pavey, G., Ahamat, G., Leutner, Franziska, Goebel, R., Knight, A., Adams, J., Hitrova, C., Barnett, J., Nachev, P., Barber, D., Chamorro-Premuzic, T., Klemmer, K., Gregorovic, M., Khan, S., & Lomas, E. (2021). Towards Algorithm Auditing: A Survey on Managing Legal, Ethical and Technological Risks of AI, ML and Associated Algorithms. *SSRN Electronic Journal.* http://dx.doi.org/10.2139/ssrn.3778998]

Mittelstadt, B.D., Allo, P., Taddeo, M., Wachter, S., & Floridi, L. (2016). The ethics of algorithms: Mapping the debate. Big Data & Society, https://doi.org/10.1177/2053951716679679

Noble, S. U. (2019). *Algorithms of oppression.* New York: NYU Press.

O'Neil, C. (2017). *Weapons of math destruction: How big data increases inequality and threatens democracy.* New York: Crown.

Ortner, S. B. (2005). Subjectivity and cultural critique. *Anthropological Theory, 5*(1).

Pangrazio, L., & Selwyn, N. (2019). "Personal data literacies": A critical literacies approach to enhancing understandings of personal digital data. *New Media and Society, 21*(2), 419–437.

Rebughini, P. (2014). Subject, subjectivity, subjectivation. *Sociopedia.Isa*.

Shilling, C. (2005). *The body in culture, technology and society*. London: Sage.

Swart, J. (2021). Experiencing algorithms: How young people understand, feel about, and engage with algorithmic news selection on social media. *Social Media and Society*, 7(2).

Wachter-Boettcher, S. (2018). *Technically wrong: Sexist apps, biased algorithms, and other threats of toxic tech*. New York: W. W. Norton and Company.

Index